Lecture Notes in Mathematics

Edited by A. Dold and B. Eckmann

Subseries: Forschungsinstitut für Mathematik, ETH Zürich

1011

Israel Michael Sigal

Scattering Theory for Many-Body Quantum Mechanical Systems –

Rigorous Results

Springer-Verlag
Berlin Heidelberg New York Tokyo 1983

Author

Israel Michael Sigal
Department of Theoretical Mathematics
The Weizmann Institute of Sciences
Rehovot 76100, Israel

AMS Subject Classifications (1980): Primary: 81 A 81, 47 A 40, 35 P 25
Secondary: 47 A 55, 35 B 40, 35 D 05

ISBN 3-540-12672-4 Springer-Verlag Berlin Heidelberg New York Tokyo
ISBN 0-387-12672-4 Springer-Verlag New York Heidelberg Berlin Tokyo

Printing and binding: Beltz Offsetdruck, Hemsbach/Bergstr.
2146/3140-543210

Contents

1. Introduction

A mathematical basis for the quantum-mechanical scattering theory was laid out in terms of the wave operators about 30 years ago. For instance the scattering matrix was expressed as a fibre of pair products of these operators. As a result, the main mathematical problem of the quantum-mechanical scattering theory was defined as the proof of existence of the wave operators and establishing their properties, the latter being (i) isometry, (ii) mutual orthogonality, and (iii) asymptotic completeless.

The existence proof was found very fast. Its different versions go back to a simple and very effective integral-of-derivative criterion of J. Cook [Co]. At the same time it was shown by J.M. Jauch [J] that the existence implies readily the isometry and mutual orthogonality. So the existence and first two qualities were finished up to certain cosmetic refinements very quickly and effectively. The asymptotic completeness however proved, from the very beginning, to be a hard nut to crack and required a certain mathematical sophistication in tackling it. Moreover, the difficulty of the problem proved to depend dramatically on the number of particles. While the two-particle systems provided a fertile ground for many authors in the field (note here important contributions by Ya. Povzner [Po], T. Ikebe [I], Kato-Kuroda [KK1,2]) and the problem was finally solved for all short-range potentials ($o(|x|^{-1})$ at infinity) by S. Agmon [A3] and V. Enss [E1] and for a large class of the long-range potentials ($O|x|^{-\alpha}$), $0 < \alpha \leq 1$, at onfinity) (for references see [RSIII], [AJS] and [Hö]), the ridges of three particle systems appeared to be difficult to penetrate. It took more than 10 years after the first classical work of L.D. Faddeev [F] on the three-body problem for the following up works to appear [CG, T2, How9, Mo1, Yaj, Yaf1]. These papers improve considerably on Faddeev's method (the elegent Hilbert-space proof of Combescure (Moulin)-Ginibre [CG] motivated by the work of R. Newton [N2] draws special attention). However, their results are very close to those of Faddeev. Namely, they prove the asymptotic completeness for three-body systems with potentials $o(|x|^{-2})$ at infinity (and with various smoothness assumptions) and under certain implicit conditions (the three-particle version of conditions (QB) and (IE) below). One of the conditions, (IE), was later dispensed with in [H2] but a high price was paid with a very involved proof.

Only recently, in yet unpublished works, the breakthrough in removing the implicit conditions and covering the potentials vanishing at infinity as $|x|^{-2}$ and slower, was achieved. In rather complex work [LS] the asymptotic completeness was proved for the short-range three-body systems under stringent but explicit conditions on the potentials. E. Mourre [Mo5] and V. Enss [E7] have proved the asymptotic completeness for essentially all short-range three-body systems (permitting for the first time the infinite number of channels) and for certain three-body long-range systems (see also S.P. Mercuriev [MJ]).

The situation with the asymptotic completeness for the N-body systems with $N \geq 4$ is much gloomier. The problem is completely solved only for a very special case of single-channel short-range systems [He, L1,2, S6,7]. In the case of the multichannel short-range systems only limited progress was made, namely, the asymptotic completeness was proven for many-body, strongly short-range (potentials are $o(|x|^{-2})$ at infinity) systems in [S2] under implicit conditions (QB) and (IE) below) which are the generalization of the implicit three-body conditions of Faddeev et al., but much more unpleasant in this case; and for four-body, strongly short-range systems in [H2] under one implicit condition (QB).

The methods used in these works fall into following three groups:

(i) Stationary method going back to Povzner [Po], Ikebe [I], Kato-Kuroda [KK1,2], [KK1,2], Faddeev [F] and developed in [A3, CG, T2, How9, Mo1, K6-8, Yaj, Yaf1, He, H2, S2-7] with important input from Kato [K3]. This method reduces the asymptotic completeness to a certain statement about the boundary values of the resolvent which is then proved by studying equations of the Fredholm type for the resolvent.

The stationary method is somewhat cumbersome, but it produces information about the behaviour of the resolvent near the continuum which contains more than mere asymptotic completeness. This information is instrumental in constructing the scattering amplitudes, scattering matrix (see section 3), etc.

(ii) The time-dependent method of Enss [E1-7, H3, Sim2, Mo2, P1,2, Yaf2,3, G, Dev, KY, MS1,2]. It provides a short-cut proof of the asymptotic completeness analyzing directly the space-time behaviour of a system in question and employing a clever decompoistion of the phase-space.

(iii) The comutator method of Kato-Lavine-Mourre [K5, L1,2, Mo3,4, PSS,FH2]. It derives certain sufficient information about the behaviour of the resolvent near the continuum using instead of the resolvent equations, the commutators of the quantum Hamiltonian with the generator of the dilation group (or related operators), localized to small spectral intervals of the Hamiltonian. (Recall that the dilation generator, A, is the symmetrized dot product of the coordinate and momentum operators, so the sign of its commutator with the quantum Hamiltonian H, (multiplied by i) shows by virtue of the Heisenberg equation $\frac{d}{dt} A(t) = e^{iHt} i[H,A] e^{-iHt}$, where $A(t) = e^{iHt} A e^{-iHt}$, whether the momentum is aligned with the coordinate or not, i.e. it describes the direction of propagation. It is not surprising therefore that the dilation generator (or related operators) and its commutators with the Hamiltonian play the basic role in the Enss method).

These notes are devoted to the mathematical foundations of the N-body scattering theory as outlined above. A small part of them describes essentially known results in the field, while the rest is concentrated on the crux of the problem; the N-body asymptotic completeness. The method we employ descends from [S2] and the main result is a modification of that of [S2]. These notes break up into two parts: an abstract scattering theory and the resolvent estimates with technical derivations of the second part carried out into the appendices. In this way the problem of different types are decoupled so we can understand them better and treat them separately. Such an approach was first introduced by Kato and Kuroda [KK1,2] in the one-particle case. For the many-body systems it was indicated by J. Howland [How9] and was developed to its present form by T. Kato [K8], M. Schechter [Schl] and I.M. Sigal [S4,5].

We now formulate the main result of these notes. First we state the assumptions:

(SR) the potentials are from the class $L^q \cap L^p_\delta(\mathbb{R}^\nu)$, where $L^p_\delta(\mathbb{R}^\nu) \equiv (1+|x|)^{-\delta/2} L^p(\mathbb{R}^\nu)$, with $\delta > 1$, $p < \nu, q > \max(2, \frac{\nu}{2})$, and are dilation analytic.

(QB) no subsystem has quasibound states at its two-cluster thresholds,

(IE) no decomposed system has embedded eigenvalues (i.e. eigenvalues embedded into its continuous spectrum).

The abbreviations in the parentheses above stand for Short-Range, Quasibound and Imbedded Eigenvalues. ν is the dimension of the physical space, we assume $\nu \geq 3$.

We demonstrate in Section 10 that conditions (IE) and (QB) while practically unverifiable, are satisfied generically. More precisely, if these conditions fail for certain pair potentials, then these potentials can be perturbed (as little as wished) in such a way that the resulting potentials obey (IE) and (QB).

The dilation-analyticity required in (SR) can be replaced by a stronger condition on the decay.

(SR') the potentials belong to the class $L^2_\delta(\mathbb{R}^\nu)$, $\delta > 2$.

The main result is

<u>Theorem.</u> Under conditions (SR) (or (SR')), (QB) and (IE) the channel wave

operators for a quantum N-body system in question exist and are asymptotically complete.

Now we explain the terms in condition (IE) and (QB). By a decomposed system we understand a system consisting of noninteracting subsystems of the original system. The notion of a quasibound state is more involved. First of all notice, that this notion makes sense only for the systems with potentials $o(|x|^{-2})$ at infinity (the dilation analyticity is unnecessary but very helpful). We say that a many-body system described by a Schrödinger operator H has a quasibound state at its two-cluster threshold λ^{α} if the equation $(H(\zeta)-\lambda^{\alpha})\psi=0$, where $\mathrm{Im}\zeta \neq 0$ and $H(\zeta)$ is the dilation-family associated with H, has a non-trivial, non-L^2 solution from the space $\psi^{\alpha}(\zeta) \otimes (\Delta_{\alpha})^{-1} L_{\delta}^{m}(\mathbb{R}^{\nu}) + \mathcal{D}(H)$, $\delta > 1$, $m < \frac{2\nu}{\nu+2}$. Here $\psi^{\alpha}(\zeta)$ is the dilation-transformed bound state for the internal motion in the channel (break-up) α, Δ_{α} is the Laplacian in the intercluster coordinates for α and $\mathbb{R}^{\nu}$ is identified with the space of those coordinates. We give such a solution the noble name of quasibound state since under small and appropriate changes of potentials it becomes a bound state and vice versa. Note that in the two-body case the dilation-analyticity is immaterial. A quasibound state is defined in this case, as a solution to $H\psi = 0$ (recall that 0 is the only (two-cluster threshold of H in this case) and the quasibound-state space becomes just $L_{\delta}^{m}(\mathbb{R}^{\nu})$, $\delta > 1$, $m < \frac{2\nu}{\nu+2}$.

To prove the theorem we use the stationary scattering theory. Hence our main effort is channeled toward the study of behaviour of the resolvent near the continuous spectrum. The latter is realized by studying equations of the Fredholm type on certain anisotropic Sobolev spaces. A slight modification of our result on the resolvent boundary values implies the analyticity of the scattering matrix for single-channel systems (see theorem 4.13 of section 4). A similar result also holds for certain diagonal channels and for (k-cluster → 2-cluster)- , (2-cluster → k-cluster)-channels in the multichannel case. The (2-cluster → 2-cluster) result belongs to Hagedorn [H1] while the results about the other cases are unpublished. (See also Balslev [B2,3] for two -and three-particle results).

Finally, we describe the content of these lecture notes. Section 2 contains some general results from the operator theory, used extensively in the main text and presented in a form fitted to the problem on hand but different from the one used in the textbooks and monographs. In section 3 we concentrate on the abstract two-space scattering theory which makes up a framework for the consequent contrete analysis. Section 4-9 are devoted to an analysis of quantum many-body systems. They start with basic definitions and end up with specific hard estimates which furnish the proof of the main result (section 8). (More details can be obtained from the table of contents). The technical developments from this part are carried out to the appendices. In section 9 we describe how to remove the assumption of dilation analyticity and in section 10 we discuss implicit conditions (IE) and (QB). The supplements collect known abstract statements frequently used in the notes. Remarks and brief reference comments are given after that.

Now a little about the history of these lecture notes. They originated in a series of lectures and seminars given by the author at Tel-Aviv University (1976), ETH, Zurich (1976-77) and Princeton University (1978), as well as in single seminars given in many different places. A preliminary version has appeared in the form of two Zurich preprints (1977-78).

The author is enormously indebted to many different people who contributed, during this long period, to his understanding of the quantum many-body problem. His special gratitude is due to E. Balslev, J.M. Combes, W. Hunziker, E. Mourre, B. Simon, and K. Yajima. The author is indebted to P.D. Milman for many discussions that contributed to the contents of section 2.

The excellent job of bringing up the difficult hand-written manuscript to the present form was done by the secretaries of the Department of Mathematics of The Weizmann Institute. My special gratitude goes to Sarah Fleigelman who did most of the first draft and to Miriam Abraham who did all of the final manuscript.

List of Symbols

$\mathcal{D}(A)$	=	the domain of an operator A.
$\mathcal{R}(A)$	=	the range of an operator A.
$\sigma(A)$	=	the spectrum of an operator A.
$\sigma_p(A)$	=	the point spectrum of an operator A = the set of all eigenvalues of A having finite multiplicities.
$\sigma_d(A)$	=	the discrete spectrum of an operator A = the set of all isolated eigenvalues of A having finite multiplicities $(\sigma_d(A) \subset \sigma_p(A))$.
$\sigma_{ess}(A)$	=	the essential spectrum of an operator $A = \sigma(A) \setminus \sigma_d(A)$.
$R_A(z)$	=	$(A-z)^{-1}$ = the resolvent of an operator A.
$\mathbb{1}$	=	an identity operator; usually it is decorated by the same indices as the space on which it acts.
$\sigma_{s.c.}(A)$	=	the singular continuous spectrum of an operator A.
$\lVert x\rVert_X$	=	the norm in a Banach space X.
$L(X,Y)$	=	the Banach space of all bounded operators from a Banach space X to a Banach space Y, $L(X) = L(X,X)$.
$L_s(X,Y)$	=	the set $L(X,Y)$ with the strong operator topology.
$\lVert A\rVert_{X\to Y}$	=	the norm in $L(X,Y)$ $(= \sup_{x\in X}(\lVert Ax\rVert_Y/\lVert x\rVert_X))$.
$\lVert A\rVert_{p\to q}$	=	the norm in $L(L^p,L^q)$.
$\lVert f\rVert_p$	=	the norm in L^p $(=[\int\lvert f\rvert^p]^{1/p}$ for $p < \infty$ and $= \sup\lvert f\rvert$ for $p = \infty)$.
$\langle f,g\rangle$	=	the inner product in L^2 $(=\int\bar{f}g)$.
$C_0^\infty(X)$	=	the space of all infinitely differentiable functions from X to $\mathbb{C}$. which have compact supports.
$C_0(\mathbb{R}^n)$	=	the space of all continuous functions from $\mathbb{R}^n$ to $\mathbb{C}$, vanishing at ∞.
$H_k(X)$	=	the Sobolev L^2- space of order k.
$A^{-1}(X)$	=	the preimage of X under an operator A $(=\{x \in \mathcal{D}(A),\ Ax \in X\})$.
$V + W$	=	the sum of two subspaces of a linear space $(=\{v+w,\ v \in V \text{ and } w \in W\})$
$V \oplus W$	=	the direct sum of two spaces $(= \{(v,w),\ v \in V \text{ and } w \in W\}$, it is also denoted $V \times W)$.

X'	=	the space dual to a Banach space X.
Ker A	=	the null space of an operator A $(=\{x \in \mathcal{D}(A),\ Ax = 0\})$.
L^2_δ	=	the weighted L^2-space $= (1+|x|^2)^{-\delta/2} L^2(dx)$.
$L^2(\Delta,X)$	=	the space of all functions, f, from a Borel interval $\Delta \subset \mathbb{R}$ to a Banach space X.s.t. $\|f(\lambda)\|_X$ is an element of $L^2(\Delta)$.
f * g	=	the convolution of functions f and g.
ImA	=	the imaginary part of an operator $A(=\frac{1}{2i}(A-A^*)$ on $\mathcal{D}(A) \cap \mathcal{D}(A^*))$.

All the Hilbert spaces figuring in this book are separable.

2. Spectral Decomposition

In this section we derive tha main theorem of theory of self-adjoint operators - the spectral theorem. We do it in a way motivated by and suitable for the abstract scattering theory. The latter will be developed in the next section.

For a given self-adjoint operator A, we introduce the operator-valued function

$$\delta_\varepsilon(A-\lambda) = \frac{1}{\pi} \operatorname{Im} R_A(\lambda + i\varepsilon), \quad \varepsilon > 0. \tag{2.1}$$

This function, an approximate δ - function of $A-\lambda$, is the central object in our approach.

Lemma 2.1. $\delta_\varepsilon(A-\lambda)$ is a positive operator-finction.

Proof. By the first resolvent equation

$$\delta_\varepsilon(A-\lambda) = \frac{\varepsilon}{\pi} R_A(\lambda-i\varepsilon) R_A(\lambda + i\varepsilon). \tag{2.2}$$

Since $R_A(\lambda-i\varepsilon) = R_A(\lambda+i\varepsilon)^*$, the lemma follows. □

Henceforth $\Delta \to \mathbb{R}$ means that Δ runs over an increasing sequence of bounded intervals whose union is $\mathbb{R}$.

Lemma 2.2. The following relation holds

$$\int \delta_\varepsilon(A-\lambda)d\lambda = 1 , \tag{2.3}$$

where the integral is understood in the weak sense or as the weak limit

$$\int_\Delta \delta_\varepsilon(A-\lambda)d\lambda \overset{w}{\to} \int \delta_\varepsilon(A-\lambda)d\lambda \quad \text{as} \quad \Delta \to \mathbb{R}. \tag{2.4}$$

Proof. We compute

$$\frac{d}{ds} \int \delta_\varepsilon(A-\lambda)d\lambda = \int A \frac{\partial}{\partial\lambda} \delta_\varepsilon(sA-\lambda)d\lambda = A\delta_\varepsilon(sA-\lambda)\Big|_{\lambda=-\infty}^{\lambda=\infty} = 0 \tag{2.5}$$

as a sesquilinear form on $\mathcal{D}(A) \times \mathcal{D}(A)$. Indeed, since $[(A-\lambda)^2+\varepsilon^2]^2 \geq \lambda^4 - 4\lambda^3 A$ we find for $v_\lambda = [(A-\lambda)^2+\varepsilon^2]^{-1}u$, $\|u\|^2 = \langle v_\lambda, [(A-\lambda)^2+\varepsilon^2]^2 v_\lambda\rangle \geq \lambda^4\|v_\lambda\|^2 - 4\lambda^3 \langle v_\lambda, Av_\lambda\rangle$. Since $\|v_\lambda\| \leq \varepsilon^{-2}\|u\|$ and $\|Av_\lambda\| \leq \varepsilon^{-2}\|Au\|$, we obtain

$$\|v_\lambda\|^2 \leq \lambda^{-4}\|u\|^2 + 4\lambda^{-1}\|v_\lambda\|\,\|Av_\lambda\| \to 0 \quad \text{as } \lambda \to \infty \quad \text{if} \quad u \in \mathcal{D}(A).$$

Since $\delta_\varepsilon(sA-\lambda)$ is continuous in s, (2.5) implies

$$\int\delta_\varepsilon(sA-\lambda)d\lambda = \int\delta_\varepsilon(-\lambda)d\lambda = \mathbb{1}. \quad \square$$

Now we introduce an <u>approximate function of A:</u>

$$f_\varepsilon(A) = \int f(\lambda)\delta_\varepsilon(A-\lambda)d\lambda \quad \text{for} \quad f \in C_0. \tag{2.6}$$

<u>Lemma 2.3.</u> The operators $f_\varepsilon(A)$ have the following properties

(i) $f_\varepsilon(A)^* = \bar{f}_\varepsilon(A)$, (ii) $\|f_\varepsilon(A)\| \leq \|f\|_\infty$ and (iii) $f_\varepsilon(A)\varphi_\varepsilon(A) - (f\varphi)_\varepsilon(A) \to 0$

as $\varepsilon \downarrow 0$ (in the operator norm).

<u>Proof.</u> (i) follows from definition (2.6). (ii) follows from the facts that $\delta_\varepsilon(A-\lambda) > 0$ and $\int\delta_\varepsilon(A-\lambda)d\lambda = \mathbb{1}$.

To prove (iii) we first use the definition $\delta_\varepsilon(A-\lambda) = (2\pi i)^{-1}[R_A(\lambda+i\varepsilon) -R_A(\lambda-i\varepsilon)]$ and the first resolvent equation

$$R_A(z)R_A(w) = (R_A(z)-R_A(w))(z-w)^{-1}$$

to transform

$$\delta_\varepsilon(A-\lambda)\delta_\varepsilon(A-s) = \frac{1}{2}\,\delta_{2\varepsilon}(\lambda-s)(\delta_\varepsilon(A-\lambda) + \delta_\varepsilon(A-s))$$

$$- (2\pi)^{-2}[R_A(\lambda+i\varepsilon)+R_A(\lambda-i\varepsilon)-R_A)s+i\varepsilon)-R_A(s-i\varepsilon)]\,\varphi_\varepsilon(\lambda-s)\ , \tag{2.7}$$

where $\varphi_\varepsilon(w) = w^{-1} - \frac{1}{2}\,(w-2i\varepsilon)^{-1} + (w+2i\varepsilon)^{-1})$. Integrating (2.7) against $f(\lambda)g(s)$ we compute

$$f_\varepsilon(A)g_\varepsilon(A) = \int f(\lambda)g(\lambda)\delta_\varepsilon(A-\lambda)d\lambda + R'_\varepsilon + R''_\varepsilon\ ,$$

where $R'_\varepsilon = \psi(A)$ with

$$\psi(\lambda) = \frac{1}{2}[f(\lambda)g_{2\varepsilon}(\lambda)-g(\lambda)) + (f_{2\varepsilon}(\lambda)-f(\lambda))g(\lambda)]$$

and $f_\varepsilon(\lambda) = \int\delta_\varepsilon(\lambda-s)f(s)ds$, the Poisson integral of f, and

$$R''_\varepsilon = (2\pi)^{-1}\varepsilon\int(R_A(\lambda+i\varepsilon) + R_A(\lambda-i\varepsilon))\alpha(\lambda)d\lambda$$

with

$$\alpha(\lambda) = [(\varphi_\varepsilon * f)(\lambda)g(\lambda) - (\varphi_\varepsilon * g)(\lambda)f(\lambda)]\varepsilon^{-1}\ ,$$

with the integral in $\lambda^{-1} * f(\lambda)$ understood in the sense of principal value.

By the property of the Poisson integral, $\|f_\varepsilon - f\|_\infty \to 0$ as $\varepsilon \downarrow 0$ for any $f \in C_0(\mathbb{R})$ (see [SW]). So $\|R'_\varepsilon\| \to 0$ by (ii). Furthermore, using that

$$|f*\varphi_\varepsilon| = |\int \frac{f(\lambda+\varepsilon x)-f(\lambda)}{x(x^2+1)} dx| \leq \varepsilon \|f'\|_\infty \int \frac{dx}{x^2+1}$$

we find

$$\|\alpha\| \leq \|f\| \, \|g'\|_\infty + \|g\| \, \|f'\|_\infty .$$

So using that $\varepsilon \int \|R_A(\lambda \pm i\varepsilon) u\|^2 d\lambda = 2\pi \|u\|^2$ by lemma 2.2 and applying the Cauchy-Schwarz inequality we obtain

$$\|R''_\varepsilon u\| \leq 2\varepsilon^{\frac{1}{2}} (2\pi)^{-1} \int \frac{dx}{x^2+1} (\|f\| \, \|g'\|_\infty + \|g\| \, \|f'\|_\infty) \|u\| .$$

Statement (ii) and the standard continuity argument complete the proof of (iii). □

Now we introduce the central notion of operator calculus.

<u>Lemma 2.4.</u> For any $f \in C_0(\mathbb{R})$, the following limit exist

$$f(A) = w - \lim_{\varepsilon \to 0} f(A). \tag{2.8}$$

<u>Proof.</u> Since $\langle \delta_\varepsilon (A-\lambda) x, x \rangle$ is a harmonic in $\lambda + i\varepsilon \in \mathbb{C}^+$, and $\int |\langle \delta_\varepsilon (A-\lambda) x, x\rangle| d\lambda \leq \|x\|^2$, it has the boundary value in the w^*-sense (see [SW]): $\int f(\lambda) \langle \delta_\varepsilon (A-\lambda) x, x \rangle d\lambda$ converges as $\varepsilon \downarrow 0$ for any $f \in C_0$. □

Lemmas 2.3 and 2.4 imply

<u>Theorem 2.5.</u> The map $f \to f(A)$ of $C_0(\mathbb{R})$ into $L(H)$ is linear and obeys (α) $f(A)^* = \bar{f}(A)$, (β) $\|f(A)\| \leq \|f\|_\infty$, ($\gamma$) $f(A)g(A) = (fg)(A)$ and (δ) $f(A) \geq 0$ whenever $f \geq 0$ (preserves the order).

To convince ourselves that (2.8) defines a usual function of A we could compute, using the same transformations as in the proof of lemma 2.3 (iii), that

$$f(A) = (A-z)^{-1} \quad \text{for} \quad f(\lambda) = (\lambda - z)^{-1}. \tag{2.9}$$

We leave this computation as an exercise to the reader.

One can extend, in a standard way, definition (2.8) to all functions f which are linear combinations of bounded functions which are pointwise limits of increasing sequences of non-negative C_0 functions. In particular, we define

$$E(\Delta,A) = \chi_\Delta(A) , \tag{2.10}$$

where χ_Δ is the characteristic function of the set Δ: $\chi_\Delta(\lambda) = 1$ if $\lambda \in \Delta$ and $= 0$ if $\lambda \notin \Delta$.

Theorem 2.6. The operator-valued function $E(\Delta,A)$ of Borel subsets of $\mathbb{R}$ has the following properties

$$E(\Delta,A)E(\Delta',A) = E(\Delta \cap \Delta',A) , \tag{2.11}$$

$$\Sigma\, E(\Delta_i,A) = E(\cup\Delta_i,A) \quad \text{if} \quad \Delta_i \cap \Delta_j = \emptyset \quad \text{for } i \neq j \tag{2.12}$$

and

$$E(\Delta,A) \overset{s}{\to} \mathbb{1} \quad \text{as} \quad \Delta \to \mathbb{R} . \tag{2.13}$$

Proof. Definition (2.10) and theorem 2.5 imply (2.11), (2.12).

To prove (2.13) we note that the operators $E(\Delta,A)$ are positive, increasing as $\Delta \to \mathbb{R}$ and $E(\Delta,A) \leq 1$. Hence $E(\Delta,A)$ converge weakly, In fact, by virtue of (2.11) they converge strongly:

$$\|(E(\Delta_n,A) - E(\Delta_m,A))\, x\, \|^2 = \|E_{\Delta_n} x\, \|^2 + \|E_{\Delta_m} x\, \|^2$$

$$- 2(E_{\Delta_n \cap \Delta_m} x,x) = (E_{\Delta_m} x,x) - (E_{\Delta_n} x,x) \to 0$$

for $\Delta_m \supset \Delta_n \to \mathbb{R}$ (remember, $L(H)$ is complete).

Let $\lim_{\Delta\to\mathbb{R}} E(\Delta,A) = P \leq \mathbb{1}$. Similarly, $\varphi_n(A) \to P$, whenever $\varphi_n \uparrow$ (identical 1), $\varphi_n \geq 0$ and $\varphi_n \in C_0$. For any $f \in C_0$ we get by theorem 2.5(β,γ):

$$f(A) = \lim_{n\to\infty} (f\varphi_n)(A) = \lim_{n\to\infty} f(A)\varphi_n(A) = f(A)P.$$

Applying this to $f(\lambda) = (\lambda-z)^{-1}$ and using (2.9), we obtain that $(A-z)^{-1} = (A-z)^{-1}P$. Since $\operatorname{Ker}(A-z)^{-1} = \{0\}$, this implies $P = \mathbb{1}$. □

Theorem 2.7. Let f be a continuous bounded function of $\mathbb{R}$. Then

$$f(A) = \int f(\lambda)\, dE(\lambda,A) , \tag{2.14}$$

where the integral is understood in a usual sense as the strong limit of integrals of finite-valued approximations of f.

Proof. The existence of the integral on the r.h.s. of (2.14) is proved in a standard way. Now we prove that quadratic forms of both sides in (2.14) are equal.

Since $\langle\delta_\varepsilon(A-\lambda)x,x\rangle$ is a harmonic function of $\lambda+i\varepsilon\in\mathbb{C}^+$ and $\int|\langle\delta_\varepsilon(A-\lambda)x,x\rangle|d\lambda\leqslant\|x\|^2$, it is a Poisson-Stieltjes integral of a finite Borel measure [SW], say $\mu(\Delta,x)$:

$$\langle\delta_\varepsilon(A-\lambda)x,x\rangle = \int\delta_\varepsilon(s-\lambda)d\mu(s,x). \tag{2.15}$$

Changing the order of integration in

$$\langle E(\Delta,A)x,x\rangle = \lim_{n\to 0}\lim_{\varepsilon\downarrow 0}\int d\lambda\, g_n(\lambda)\ \ [\int\delta_\varepsilon(s-\lambda)d\mu(s,x)],$$

where $g_n\geqslant 0$ is an increasing sequence converging to χ_Δ, passing to the limit under the integral sign (by the Lebesque dominated convergence theorem) and using convergence properties of the Poisson integral [SW], we obtain

$$\langle E(\Delta,A)x,x\rangle = \mu(\Delta,x). \tag{2.16}$$

Finally, changing the order of integration in

$$\langle f(A)x,x\rangle = \lim_{\varepsilon\downarrow 0}\int f(\lambda)[\int\delta_\varepsilon(s-\lambda)d\mu(s,x)]d\lambda$$

and passing to limit under the integral sign we arrive at

$$\langle f(A)x,x\rangle = \int f(s)d\mu(s,x), \tag{2.17}$$

which, in virtue of (2.16), implies (2.14). □

The operator calculus developed above can be extended, in a standard way, to unbounded functions. We omit here this step, presenting instead a special but very important case $f(\lambda)=\lambda$:

Theorem 2.8. A self-adjoint operator A is representable on $\mathcal{D}(A)$ as

$$A = \int\lambda dE(\lambda,A)$$

(the interpretation of the integral on the r.h.s. is given in theorem 2.7).

Proof. Applying theorem 2.7, to $f_n(\lambda)=\lambda g_n(\lambda)$, where $g_n\geqslant 0$ is an increasing sequence converging to χ_Δ, we obtain that $\int_\Delta\lambda dE(\lambda,A) = AE(\Delta,A)$ for any bounded interval. By (2.13), $AE(\Delta,A)x\to Ax$ as $\Delta\to\mathbb{R}$ for $x\in\mathcal{D}(A)$. □

Theorem 2.8. is the central theorem of the spectral theory of self-adjoint operators, it is called the spectral theorem (sometimes, an extension of theorem 2.5 to bounded functions is called the spectral theorem).

Remark 2.9. Instead of (2.9) one can use in the proof of theorem 2.6, eqn. (2.15).

Let $E^{(\varepsilon)}(\Delta,A) = \int_\Delta \delta_\varepsilon(A-\lambda)d\lambda$. Certain properties of this family, described in the following two statements, are needed in the next section.

Lemma 2.10. For any Borel $\Delta, E^{(\varepsilon)}(\Delta,A)$ is strongly continuous in $\varepsilon \geq 0$. $E^{(\varepsilon)}(\Delta,A) \overset{s}{\to} E(\Delta,A)$ as $\varepsilon \downarrow 0$.

Proof. Eqn.(2.15) implies that $E^{(\varepsilon)}(\Delta,A) = \chi_\Delta^{(\varepsilon)}(A)$, where $\chi_\Delta^{(\varepsilon)}(\lambda) = \int_\Delta \delta_\varepsilon(s-\lambda)ds$.

Since $|\chi_\Delta^{(\varepsilon)}(\lambda)| \leq 1$ and $\chi_\Delta^{(\varepsilon)}(\lambda) \to \chi_\Delta(\lambda)$ as $\varepsilon \downarrow 0$ for each $\lambda \in \mathbb{R}$, we have by the Lebesgue convergence theorem and eqn.(2.17), that $E^{(\varepsilon)}(\Delta,A) \overset{w}{\to} E(\Delta,A)$. Similarly, we have: $E^{(\varepsilon)}(\Delta,A)^2 \overset{w}{\to} E(\Delta,A)$. This together with (2.11) implies that $E^{(\varepsilon)}(\Delta,A) \overset{s}{\to} E(\Delta,A)$ for any Borel Δ. □

Lemma 2.11. Let a sequence of Borel subsets Δ_n satisfy $\Delta_n \cap [-a_n,a_n] = \emptyset$, where $a_n \to \infty$ as $n \to \infty$. Then $E^{(\varepsilon)}(\Delta_n,A) \overset{s}{\to} 0$ as $n \to \infty$, uniformly in $\varepsilon \in [0,1]$.

Proof. Again we use the representation $\langle E^{(\varepsilon)}(\Delta)^2 x,x\rangle = \int \chi_\Delta^{(\varepsilon)}(\lambda)^2 d\mu(\lambda,x)$ and the fact that $\chi_\Delta^{(\varepsilon)}(\lambda)^2 \leq 1$ and $\chi_{\Delta_n}^{(\varepsilon)}(\lambda)^2 \to 0$ as $n \to \infty$ for each $\lambda \in \mathbb{R}$ uniformly in ε, to obtain that $E^{(\varepsilon)}(\Delta_n,A)^2 \overset{w}{\to} 0$ as $n \to \infty$ uniformly in ε. This implies that $E^{(\varepsilon)}(\Delta_n,A) \to 0$, strongly, as $n \to \infty$, uniformly in ε. □

3. Two-Space Scattering Theory

The abstract scattering theory deals with results which are common for many different scattering systems and are independent of a detail structure of the operators involved. It compares the long time behaviour of (one-parameter) evolution groups of different self-adjoint operators (the time-dependent theory). It will be shown below that this problem is closely related to the problem of similarity of two self-adjoint operators. The application to the multichannel systems requires considering such operators on different Hilbert spaces. In this section we show how certain rather general conditions on a pair of self-adjoint operators distinguish the two-space scattering theory from the general theory of self-adjoint operators.

a. Time-dependent theory

Let $\mathcal{H}$ and $\hat{\mathcal{H}}$ be Hilbert spaces and H and $\hat{H}$ self-adjoint operators on $\mathcal{H}$ and $\hat{\mathcal{H}}$, respectively. Let furthermore J be a bounded operator from $\hat{\mathcal{H}}$ into $\mathcal{H}$.

Definition 3.1. The strong limits

$$W^{\pm} = s - \lim_{t\to\pm\infty} e^{iHt} J e^{-i\hat{H}t}, \tag{3.1}$$

if they exist on the absolute continuous subspace for $\hat{H}$, are called the wave operators for the triple $(H,\hat{H},J)$. When it is clear which J is chosen, we call $W^{\pm}$ the wave operators for the pair $(H,\hat{H})$.

In order to simplify notations, we assume in the sequel that the operator $\hat{H}$ is absolutely continuous. This condition is satisfied in our applications.

Operators J in (3.1) are called identications. Two operators J and J' lead to the same wave operators if and only if

$$s - \lim_{|t|\to\infty} (J-J')e^{-i\hat{H}t} = 0 .$$

Identifications satisfying (3.2) will be called equivalent.

Following W. Hunziker [Hu2] and T. Kato [K4] we assume that

$$\lim_{|t|\to\infty} \| Je^{-i\hat{H}t}\hat{f} \| = \|\hat{f}\|, \quad \hat{f} \in \hat{\mathcal{H}}, \tag{3.3}$$

i.e. J is asymptotically isometric. This condition is always fulfilled in the concrete case and is natural to the kind of problems we are studying in the abstract setting.

Denote by $\mathcal{H}_{a.c.}(H)$ the subspace of the absolute continuous spectrum of H and by E_p the eigenprojection on the point-spectrum subspace of H.

Theorem 3.2. (Kato-Hunziker). Let $W^{\pm}$ exist. Then

(i) $W^{\pm}$ are intertwinning for H and $\hat{H}$:

$$HW^{\pm} = W^{\pm}\hat{H} , \tag{3.4}$$

(ii) $\mathcal{R}(W^{\pm}) \subset \mathcal{H}_{ac}(H) \subset \mathcal{R}(\mathbb{1}-E_p)$.

If, in addition, eqn (3.3) holds, then

(iii) $W^{\pm}$ are isometries:

$$W^{\pm *}W^{\pm} = \mathbb{1} . \tag{3.5}$$

Proof. Intertwinning property of $W^{\pm}$ follows readily from (3.1). Indeed, we prove by change of the variable $t \rightarrow r = t + s$, that $e^{iHs}W^{\pm} = W^{\pm}e^{i\hat{H}s}$. The latter is equivalent to (3.4). (ii) follows from $\exp(-i\hat{H}t) \overset{w}{\rightarrow} 0$ as $|t| \rightarrow \infty$, which is true for any absolute continuous $\hat{H}$. (iii) is straightforward. □

Corollary 3.3. (i) $W^{\pm}W^{\pm *}$ are projections on $\mathcal{R}(W^{\pm})$. (ii) H and $\hat{H}$ are unitary equivalent on $\mathcal{R}(W^{\pm})$.

Definition 3.4. $W^{\pm}$ are said to be (asymptotically) complete iff $\mathcal{R}(W^{\pm}) = \mathcal{R}(\mathbb{1} - E_p)$, i.e.

$$W^{\pm}W^{\pm *} = \mathbb{1} - E_p . \tag{3.6}$$

Note that in this case $\sigma_{s.c.}(H) = \phi$.

From now on we assume that J maps $\mathcal{D}(\hat{H})$ into $\mathcal{D}(H)$, so that the operator $I = HJ - J\hat{H}$ is defined from $\mathcal{D}(\hat{H})$ to $\mathcal{H}$. Note that the operator I plays the role of $V = H - T$ of the one-space theory. If $J = \mathbb{1}$ and $\hat{H} = T$, then $I = V = H - T$.

The next theorem uses the following representation of $W^{\pm}$ which is obtained by the standard trick of integration of derivative $(\frac{d}{dt} e^{iHt}Je^{-i\hat{H}t} = ie^{iH}Ie^{-i\hat{H}t})$:

$$W^{\pm} = J - \int_0^{\pm\infty} e^{iHt}Ie^{-i\hat{H}t}dt , \tag{3.7}$$

where the integral is understood in the strong sense.

Theorem 3.5. (Cook-Kuroda). Assume there exist a set $\mathcal{D} \subset \mathcal{D}(\hat{H})$, dense in $\hat{\mathcal{H}}$ and a number $T > 0$ so that $Ie^{-i\hat{H}t}u \in L^1([\pm T, \pm\infty], \mathcal{H})$ for any $u \in \mathcal{D}$. Then $W^{\pm}$ exist.

Proof. Straightforward from (3.6) by writting $\int_0^{\pm\infty} = \int_0^{\pm T} + \int_{\pm T}^{\pm\infty}$ and using that $\mathcal{D}(\hat{H})$ is invariant under $\exp(-i\hat{H}t)$ and $I : \mathcal{D}(\hat{H}) \to \mathcal{H}$. □

Definition 3.6. The scattering operator for the triple (H,H,J) is

$$S = W^{+*}W^{-} . \tag{3.8}$$

If $\mathcal{R}(W^{+}) = \mathcal{R}(W^{-})$ then S is unitary:

$$SS^{*} = S^{*}S = \hat{1} \tag{3.9}$$

so the asymtotic completeness implies the unitarity of S.

The intertwinning property of $W^{\pm}$ implies that S commutes with $\hat{H}$:

$$[S,\hat{H}] = 0. \tag{3.10}$$

Let $\int^{\oplus}\hat{\mathcal{H}}_{\lambda}d\lambda$ be a fibre direct integral with respect to $\hat{H}$ and let Π be a unitary operator from $\hat{\mathcal{H}}$ to $\int^{\oplus}\hat{\mathcal{H}}_{\lambda}d\lambda$. Then, because of (3.10), the operator S is decomposable:

$$\Pi S \Pi^{*} = \int^{\oplus} S(\lambda)d\lambda, \tag{3.11}$$

where $S(\lambda)$ acts on $\hat{\mathcal{H}}_{\lambda}$. The operator valued function $S(\lambda)$ is called the scattering (S-) matrix.

b. Stationary theory

The wave operators can be also expressed in a time-independent (or stationary) form, in terms of the resolvents of H and $\hat{H}$. The advantage of the stationary form lies in the fact that the resolvents lend easier themselves to the study than evolution operators. The transformation from the time-dependent form to the stationary one is done by using

Lemma 3.7. (Generalized Abel's theorem). Let f be a continuous vector-function form $\mathbb{R}$ to $\mathcal{H}$ and let $\lim_{t\to\infty} f(t)$ exist. Then

$$\text{Abel} - \lim_{t\to\infty} f(t) \equiv \lim_{\varepsilon\downarrow 0} \varepsilon\int_0^{\infty} f(t)e^{-\varepsilon t}dt$$

exists and equals $\lim_{t\to\infty} f(t)$.

Another advantage of the stationary form is that it admits a <u>local formulation</u> with which we begin. Let $R(z)$ and $\hat{R}(z)$ be the resolvents of H and $\hat{H}$, respectively. For any Borel $\Delta \subset \mathbb{R}$ and $\varepsilon \in \mathbb{R}^{\pm}$ we introduce the bounded operator

$$W^{(\varepsilon)}(\Delta) = \frac{\varepsilon}{\pi} \int_{\Delta} R(\lambda-i\varepsilon) J \hat{R}(\lambda+i\varepsilon)\, d\lambda. \tag{3.12}$$

That the r.h.s. is well defined even for unbounded Δ follows from the estimate

$$\int_{\Delta} |\langle J\hat{R}(\lambda+i\varepsilon)u \,,\; R(\lambda+i\varepsilon)v\rangle|\, d\lambda$$

$$\leq \|J\| \{\varepsilon\int_{\Delta} \|R(\lambda+i\varepsilon)v\|^2 d\lambda\}^{1/2} \{\varepsilon\int_{\Delta} \|R(\lambda+i\varepsilon)u\|^2 d\lambda\}^{1/2} \tag{3.13}$$

and the inequality

$$\varepsilon\int_{\Delta} \|R(\lambda+i\varepsilon)u\|^2 d\lambda \leq \pi\|u\|^2 \tag{3.14}$$

which follows from (2.3).

<u>Definition 3.8.</u> Let Δ be a Borel subset of $\mathbb{R}$, the strong limits

$$W^{\pm}(\Delta) = \operatorname*{s-lim}_{\varepsilon\to\pm 0} W^{(\varepsilon)}(\Delta) \tag{3.15}$$

if they exist, are called <u>local stationary wave operators</u> (corresponding to Δ).

$W^{\pm}(\mathbb{R})$ are called (global) stationary wave operators and are denoted by the same symbols, $W^{\pm}$, as the nonstationary ones. We also denote $W^{(\varepsilon)} = W^{(\varepsilon)}(\mathbb{R})$. This should not cause a confusion because we have

<u>Proposition 3.9.</u> If wave operators (3.1) exist then so do the stationary wave operators. Moreover, both definitions lead to the same operators.

<u>Proof.</u> By the generalized Abel theorem, if strong limits (3.1) exist, then the Abel limits exist as well and are equal to the strong ones:

$$(3.1) = \lim_{\varepsilon\to\pm 0} \varepsilon\int_0^{\pm\infty} e^{iHt} J e^{-i\hat{H}t} e^{-\varepsilon t}\, dt. \tag{3.16}$$

<u>Lemma 3.10.</u> The following equality holds

$$W^{(\varepsilon)} = 2\varepsilon \int_0^{\pm\infty} e^{-2\varepsilon t} e^{iHt} J e^{-i\hat{H}t}\, dt. \tag{3.17}$$

<u>Proof.</u> Eqn (3.17) follows from the well-known formula

$$(A-z)^{-1} = i\int_0^{\pm\infty} e^{-iAt} e^{+izt}\, dt, \qquad \operatorname{Im} z \in \mathbb{R}^{\pm}, \tag{3.18}$$

valid for any self-adjoint operator A, and the vector-function Plancherel theorem applied to $\langle W^{(\varepsilon)}u,v\rangle$. □

This lemma together with eqn (3.16) completes the proof of prop. 3.9. □
Let $E(\Delta)$ and $\hat{E}(\Delta)$ be the spectral projections for H and $\hat{H}$, respectively (see thm. 2.6). Note that the local wave operators can be obtained from the global ones as

$W^{\pm}(\Delta) = W^{\pm}\hat{E}(\Delta)$.

The proof of this equation is simple. We omit it here since the relation is not used in this book. In our approach we study $W^{\pm}(\Delta)$ and then recover the needed information about $W^{\pm}$ from them. To the last end we use

<u>Lemma 3.11.</u> Let $W^{\pm}(\Delta)$ exist for all Δ's from a directed sequence $\Phi = \{\Delta_i\}$ of Borel subsets of IR). Let, furthermore, for any $\Delta, \Delta' \in \Phi$, $W^{\pm}(\Delta)^* W^{\pm}(\Delta') = \hat{E}(\Delta \cap \Delta')/W^{\pm}(\Delta)W^{\pm}(\Delta')^* = E(\Delta \cap \Delta')$. Then $W^{\pm}(I) = s-\lim_{\Delta\to I} W^{\pm}(\Delta)/W^{\pm}(I)^* = s\text{-}\lim_{\Delta\to I} W^{\pm}(\Delta)^*$, where $I = \cup_{\Delta\in\phi}\Delta$, exist. Moreover, if the Lebesgue measure of IR/I is zero, then $W^{\pm}(I) = W^{\pm}(IR)$, the global wave operators.

<u>Proof.</u> The first statement follows from the following simple proposition:
$A_n^* A_m$ converge strongly (resp. uniformly) as $n,m \to \infty$ implies
A_n converges strongly (resp. uniformly) as $n \to \infty$.

To prove the second statement we show that $W^{(\varepsilon)}(\Delta)$ converges to $W(I)$ as $\Delta \uparrow I$ uniformly in $\varepsilon \in IR^{\pm}$ (the similar statement holds also for $W^{(\varepsilon)}(\Delta)^*$). To this end we use that $W^{(\varepsilon)}(I) - W^{(\varepsilon)}(\Delta) = W^{(\varepsilon)}(I\setminus\Delta)$ and hence by (3.13) and (3.14),
$|\langle(W^{(\varepsilon)}(I) - W^{(\varepsilon)}(\Delta))u,v\rangle| \leq \|v\|(\langle\hat{E}^{(\varepsilon)}(I\setminus\Delta)u,u\rangle)^{\frac{1}{2}}$, where $\hat{E}^{(\varepsilon)}(\Delta) = \int_\Delta \delta_\varepsilon(\hat{H}-\lambda)d\lambda$.

The r.h.s. of the latter inequality converges to 0 as $\Delta \uparrow I$, uniformly in ε by lemma 2.11. The uniform convergence implies that the limits in Δ and ε can be interchanged in the definition of $W^{\pm}(I)$ which yields that $W^{\pm}(I) = W^{\pm}(IR)$, provided that the measure of $IR\setminus I$ is zero. □

The condition of asymptotic isometry of $Je^{-i\hat{H}t}$ in the stationary, local case can be written as

$$\lim_{|\varepsilon|\to 0} \frac{|\varepsilon|}{\pi} \int_\Delta \|J\hat{R}(\lambda+i\varepsilon)\hat{u}\|^2 d\lambda = \|\hat{E}(\Delta)\hat{u}\|^2 \quad \forall\, \hat{u} \in \hat{\mathcal{H}} . \tag{3.19}$$

<u>Lemma 3.12.</u> If $\|J\| \leq 1$ then condition (3.19) is equivalent to the condition

$$s - \lim \frac{|\varepsilon|}{\pi}\int_\Delta \hat{R}(\lambda-i\varepsilon)J^*JR(\lambda+i\varepsilon)d\lambda = \hat{E}(\Delta). \tag{SLAI}$$

<u>Proof.</u> The lemma is derived from the following simple statement (P is a projection):

$A_n \overset{w}{\to} P$, $\lim \|A_n f\| \leq \|Pf\| \Rightarrow A_n \overset{s}{\to} P$ and the relations $\frac{\varepsilon}{\pi}\int_\Delta \|J\hat{R}(\lambda+i\varepsilon)\hat{u}\|^2 d\lambda$

$\leq \|J\|^2 \|\hat{E}^{(\varepsilon)}(\Delta)\hat{u}\|^2$ and $\|\hat{E}^{(\varepsilon)}(\Delta)\hat{u}\| \to \|\hat{E}(\Delta)\hat{u}\|$ as $\varepsilon \downarrow 0$ (see lemma 2.10). □

The following theorem states sufficient conditions on the resolvents of H and $\hat{H}$ in order for the stationary $W^{\pm}$ to be complete. These conditions roughly mean that the resolvents of both operators are <u>proportional up to an operator-valued function</u> on $\mathbb{C}\setminus\mathbb{R}$, which has <u>strong boundary values on $\mathbb{R}$</u> when considered between appropriate Banach spaces.

<u>Definition.</u> Let T be a self-adjoint operator on a Hilbert space $\mathcal{H}$. A Banach space X will be called T-smooth if (i) $X \cap \mathcal{H}$ is dense in X and (ii) $\delta_\varepsilon(T-\lambda)$ extends to a family of operators from X to its dual X', uniformly bounded in $\varepsilon \in \mathbb{R}^+$ and $\lambda \in \mathbb{R}$.

<u>Theorem 3.13.</u> Let $\hat{H}$ and J satisfy (SLAI). Let the resolvents of H and $\hat{H}$ be connected by the equation

$$(\mathbb{1} - E_p)R(z) = J\hat{R}(z)Q(z), \tag{3.20}$$

where $Q(z)$, $z \in \mathbb{C}\setminus\mathbb{R}$, is a family of operators from $\mathcal{H}$ to $\hat{\mathcal{H}}$. Assume also that there exist a dense subset $Y \subset (\mathbb{1}-E_p)\mathcal{H}$, a collection Φ of Borel subsets of $\mathbb{R}$ and a Banach space $\hat{X}$ such that

(i) $\hat{X}$ is $\hat{H}$-smooth,

(ii) For any $f \in Y$, $\varepsilon \in \mathbb{R}$, $\Delta \in \Phi$; $Q(\cdot+i\varepsilon)f \in L^2(\Delta,\hat{X})$ and has strong limits in $L^2(\Delta,\hat{X})$ as $\varepsilon \to \pm 0$: $\|(Q(\cdot+i\varepsilon) - Q(\cdot+i\varepsilon'))f\|_{L^2(\Delta,\hat{X})} \to 0 (\varepsilon,\varepsilon' \to \pm 0)$.

Then (a) $\sigma_{s.c.}(H) \cap (\bigcup_{\Delta\in\Phi} \Delta) = \emptyset$, (b) $W^{\pm}(\Delta)^* = s - \lim_{\varepsilon\to\pm 0} W^{(\varepsilon)}(\Delta)^*$, exist for any $\Delta \in \Phi$ and equal $\lim W_1^{(\varepsilon)}(\Delta)$, where $W_1^{(\varepsilon)}(\Delta)^* = \int_\Delta \delta_\varepsilon(\hat{H}-\lambda)Q(\lambda+i\varepsilon)d\lambda$

(c) $W^{\pm}(\Delta)W^{\pm}(\Delta')^* = E(\Delta \cap \Delta')(\mathbb{1} - E_p)$, $\Delta,\Delta' \in \Phi$.

Before proceeding to the proof of the theorem we will demonstrate a few preliminary results for the operators $\hat{H}$ and J.

<u>Lemma 3.14.</u> Let $\hat{X}$ be a Banach space such that $\hat{X} \cap \hat{\mathcal{H}}$ is dense and $\varepsilon|\int_\Delta \|\hat{R}(\lambda+i\varepsilon)x(\lambda)\|^2 d\lambda \leq M\|x\|^2_{L^2(\Delta,\hat{X})}$ (i.e. $\delta_\varepsilon(\hat{H}-\lambda)$ is bounded from $L^2(\Delta,\hat{X})$ to its dual $L^2(\Delta,\hat{X}')$ uniformly in $\varepsilon \in \mathbb{R}^{\pm}$). Then for any $x,y \in L^2(\Delta,\hat{X})$,

$$\int_\Delta d\lambda\, \delta_\varepsilon(\hat{H}-\lambda)x(\lambda) \text{ is a Cauchy sequence in } \varepsilon \text{ in } \hat{\mathcal{H}} \tag{3.21}$$

and

$$\int_{\Delta\times\Delta'} \langle\delta_\varepsilon(\hat{H}-\lambda)\delta_\varepsilon(\hat{H}-\nu)x(\nu),y(\lambda)\rangle d\lambda d\nu - \int_{\Delta\cap\Delta'} \langle\delta_\varepsilon(\hat{H}-\lambda)x(\lambda),y(\lambda)\rangle d\lambda \to 0 \quad (\varepsilon \to 0). \tag{3.22}$$

Proof. Since by the condition of the lemma, $\int \|\delta_\varepsilon(\hat{H}-\lambda)^{\frac{1}{2}} x(\lambda)\|^2 d\lambda \leq M\|x\|^2_{L^2(\Delta,\hat{X})}$, and since by the other condition, $\mathrm{Span}\{f\cdot x,\ f\in L^2(\Delta),\ x\in\hat{X}\cap\hat{H}\}$ is dense in $L^2(\Delta,\hat{X})$, it suffices to show that $\int_\Delta \delta_\varepsilon(\hat{H}-\lambda)x d\lambda = \hat{E}^{(\varepsilon)}(\Delta)x$ is Cauchy in $\hat{H}$. The latter results from lemma 2.10. Eqn. (3.22) is proved exactly in the same way by reducing it to (iii) of lemma 2.3. □

Lemma 3.15. If in addition to the conditions of lemma 3.14, (SLAI) holds as well, then as $\varepsilon\to\pm 0$,

$$|\varepsilon|\,\|\int_\Delta \hat{R}(\lambda-i\varepsilon)(J^*J-\hat{1\!\!1})\hat{R}(\lambda+i\varepsilon)x(\lambda)d\lambda\| \to 0 \tag{3.23}$$

and

$$|\varepsilon|\int_\Delta \langle x(\lambda),\ \hat{R}(\lambda-i\varepsilon)(J^*J-\hat{1\!\!1})\hat{R}(\lambda+i\varepsilon)y(\lambda)\rangle d\lambda \to 0. \tag{3.24}$$

Proof. Using the same approximation argument as in the proof of lemma 3.14 we reduce both relations to (SLAI) (note that in order to estimate (3.23) we estimate $|\varepsilon|\int_\Delta \langle (J^*J-1\!\!1)\hat{R}(\lambda+i\varepsilon)x(\lambda),\ \hat{R}(\lambda+i\varepsilon)v\rangle d\lambda$ for any $v\in H$ using (3.14)). □

Proof of theorem 3.13. (a) A simple derivation shows that the existence of $W^\pm(\Delta)^*$ implies the intertwining property: $\hat{E}(\Delta)W^\pm(\Delta)^* = W^\pm(\Delta)E(\Delta)$. This yields that $R(W^\pm)\subset H_{a.c}(H)$. Thus (b) and (c) imply (a).

(b) $W_1^{(\varepsilon)}(\Delta)^*$ is bounded as follows from

$$|\langle W_1^{(\varepsilon)}(\Delta)^*u,v\rangle| \leq \{\int_\Delta\|\delta_\varepsilon(\hat{H}-\lambda)\ v\|^2 d\lambda\}^{\frac{1}{2}}\{\int_\Delta\|\delta_\varepsilon(\hat{H}-\lambda)^{\frac{1}{2}}Q(\lambda+i\varepsilon)u\|^2 d\lambda\}^{\frac{1}{2}}$$

$$\leq M\,\|v\|\,\{\int_\Delta\|Q(\lambda+i\varepsilon)u\|_X^2 d\lambda\}^{\frac{1}{2}}$$

$W_1^{(\varepsilon)}(\Delta)^*$ converges strongly as $\varepsilon\to\pm 0$:

$$\|W_1^{(\varepsilon')}(\Delta)^*u - W_1^{(\varepsilon)}(\Delta)^*u\| \leq 2\int\|(\delta_{\varepsilon'}(\hat{H}-\lambda)^{\frac{1}{2}} - \delta_\varepsilon(\hat{H}-\lambda)^{\frac{1}{2}})Q(\lambda+i\varepsilon)u\|^2 d\lambda)^{\frac{1}{2}}$$

$$+ \{\int_\Delta\|\delta_\varepsilon(\hat{H}-\lambda)(Q(\lambda+i\varepsilon') - Q(\lambda+i\varepsilon))u\|^2 d\lambda\}^{\frac{1}{2}}.$$

The right-hand side converges to 0 as $\varepsilon,\varepsilon'\to\pm 0$ by (i), (ii) and (3.21).

$W^{(\varepsilon)}(\Delta)^*$ converges strongly as well, since

$$W^{(\varepsilon')}(\Delta)^* - W_1^{(\varepsilon)}(\Delta)^* = \frac{\varepsilon}{\pi}\int_\Delta\hat{R}(\lambda+i\varepsilon)(J^*J-1)\hat{R}(\lambda+i\varepsilon)Q(\lambda+i\varepsilon)d\lambda \to 0$$

strongly as $\varepsilon\to\pm 0$ by (3.23).

(c) Using (3.22),

$$W_1^{(\varepsilon)}(\Delta)W_1^{(\varepsilon)}(\Delta')^* = \int_{\Delta\cap\Delta'} Q(\lambda+i\varepsilon)^*\delta_\varepsilon(\hat{H}-\lambda)Q(\lambda+i\varepsilon)d\lambda + O(1)$$

weakly as $\varepsilon \to \pm 0$. Using eqn (3.20) and (3.24)

$$\int_\Delta \delta_\varepsilon(H-\lambda)(\mathbb{1} - E_p)d\lambda = \int_\Delta Q(\lambda+i\varepsilon)^*\delta_\varepsilon(\hat{H}-\lambda)Q(\lambda+i\varepsilon)d\lambda + O(|\varepsilon|^0)$$

as a sesquilinear from on Y. Comparing these two equations we arrive at the desired relation. □

Corollary 3.16. Under the conditions of the theorem the stationary global wave operators exist and are complete in the sense that $W^\pm W^{\pm *} = E(I)$, where $I = \bigcup_{\Delta\in\phi} \Delta$.

This statement follows from lemma 3.11 and theorem 3.13.

Remark. Strengthening somewhat condition (ii) and adding a new condition on the operator $(H-z)J(\hat{H}-z)^{-1}$ we can prove that $W^\pm(\Delta)$ exist as strong limits of $W^{(\varepsilon)}(\Delta)$ and satisfy $W^\pm(\Delta)^*W^\pm(\Delta') = \hat{E}(\Delta \cap \Delta')$. However, for the self-adjoint case this result is of a little interest, since the condition of existence and therefore isometry of the nonstationary, global wave operators can be checked in much easier way than our conditions.

c. An example of the X-space

Our definition of a relatively smooth Banach space is a natural generalization of Kato's definition [K3,RS1V] of a relatively smooth operator. Such operators generate also the most important example of relatively smooth Banach spaces. Below we present one of Kato's equivalent characterizations of a relatively smooth operator as a definition.

Definition 3.17. (Kato). Let T be a self-adjoint operator and A, a closed operator on $\mathcal{H}$. A is called T-smooth iff for each $u \in \mathcal{H}$, $e^{itT}u \in \mathcal{D}(A)$ for almost every $t \in \mathbb{R}$ and

$$\int_{-\infty}^{\infty} \|Ae^{-itT}u\|^2 dt \leq C\|u\|^2$$

for some constant $C < \infty$.

Below we will need the following

Kato's Lemma. Let A be a closed operator and let T be self-adjoint. Then A is T-smooth iff $A\delta_\varepsilon(T-\lambda)A^*$ is uniformly bounded in $\lambda \in \mathbb{R}$ and $\varepsilon > 0$.

Proof. See supplement I, lemma SI.11. □

If A is an operator from a Banach space H to K, then AH denotes the completion of $R(A)$ in the norm $\|u\| = \inf_{h, Ah=u} \|h\|$. The Banach space AH is isometrically isomorphic to the quatient space $H/\mathrm{Ker}\, A$. We also define $\Sigma A_i H$ as $j(\oplus A_i H)$, where j maps $\oplus K$ to K as $j(\oplus k_i) = \Sigma k_i$.

Lemma 3.18. Let the operators A_i on H be T-smooth and define $X = \Sigma A_i^* H$. Then X is T-smooth. If, in addition, $A_j \delta_\varepsilon (T-\lambda) A_i^*$ is strongly continuous on H as $\varepsilon \to 0$ then so is $\delta_\varepsilon(T-\lambda)$ from X to X'.

Proof. $X \cap H$ dense in X by the definition of X as the completion of $\Sigma R(A_i)$ in the norm of X. By Kato's lemma A_i is T-smooth if and only if $A_i \delta_\varepsilon (T-\lambda) A_i^*$ is uniformly in $\varepsilon \in \mathbb{R}^+$ and $\lambda \in \mathbb{R}$ bounded on H. This implies that

$$\|A_j \delta_\varepsilon (T-\lambda) A_i^*\|^2 \leq \|A_j \delta_\varepsilon (T-\lambda) A_j^*\| \, \|A_i \delta_\varepsilon (T-\lambda) A_i^*\|$$

is uniformly bounded on H for any i and j. The latter implies the first statement. The proof of the second statement is straightforward. □

Remark 3.19. Another proof of the first part of the lemma would be to use Kato definition 3.17 of the relatively smooth operator, which implies readily that ΣA_i is T-smooth. Then one uses the equivalence of the norms in $\Sigma A_i^* H$ and $(\Sigma A_i^*) H$.

d. Scattering matrix. Single-space case

We consider first the single-space case since it is much simpler and the proofs can be done differently. So let $\hat{H} = H$, $J = 1\!\!1$ and denote $\hat{H} = T$. Let $\int^\oplus H_\lambda d\lambda$ be the fiber direct integral associated with T defined on H. Recall, that Π is a unitary operator from H to $\int^\oplus H_\lambda d\lambda$. It can be written as $\Pi = \{\Pi_\lambda\}$, where Π_λ are operators from H to H_λ s.t. $\|\Pi_\lambda u\|_{H_\lambda} \in L^2(\mathbb{R})$ and $\int \|\Pi_\lambda u\|^2_{H_\lambda} d\lambda = \|u\|^2$ for $u \in H$. We introduce the T-operator $T(z) = V - VR(z)V$, where $V = H - T$.

Theorem 3.20. Let there exist a Banach space X such that (α) $H - T$ is bounded from the dual X' to X, (β) X is T-smooth, (γ) $(H-T)R(z)$, $z \in \mathbb{C} / \mathbb{R}$, can be extended to a family of bounded operators on X which has strong boundary values on $\mathbb{R} / \sigma_p(H)$. Then $T_\lambda \equiv \mathrm{s}-\lim_{\varepsilon \downarrow 0} \Pi_\lambda T(\lambda + i\varepsilon) \Pi_\lambda^*$ exists as a measurable $L(H_\lambda)$-

valued function uniformly bounded in $\lambda \in \mathbb{R}$ and $S(\lambda) = \mathbb{1}_\lambda + 2\pi i T_\lambda$.

Proof. We begin with

Lemma 3.21. Let a Banach space X be T-smooth. Then Π_λ extends to a family of operators from X to H_λ, uniformly bounded in $\lambda \in \mathbb{R}$. If, in addition, $\delta_\varepsilon(T-\lambda)$ has weakly continuous boundary values as $\varepsilon \downarrow 0$, then $\|\Pi_\lambda u\|_{H_\lambda}$ is continuous in λ.

Proof. First of all for any $u \in H$, $\|\Pi_\lambda u\|_{H_\lambda} \in L^2(\mathbb{R})$. We write $\langle\delta_\varepsilon(T-\lambda)u,u\rangle = \int\delta_\varepsilon(s-\lambda)\|\Pi_s u\|^2_{H_s}\, ds$, a Poisson integral of $\|\Pi_s u\|^2_{H_s}$. On the other hand, since $\langle\delta_\varepsilon(T-\lambda)u,u\rangle$ is a harmonic, positive, uniformly bounded in $\lambda \in \mathbb{R}$ and $\varepsilon \in \mathbb{R}^+$, it is a Poisson integral of a uniformly bounded function, to which it converges pointwise as $\varepsilon \downarrow 0$ (see [SW]). The proof now follows from the uniqueness of the Poisson integral. □

Lemma 3.22. If $W^\pm$ exist then

$$S - \mathbb{1} = 2\pi i \lim_{\varepsilon\downarrow 0}\int\delta_\varepsilon(T-\lambda)W^{+*}V\delta_\varepsilon(T-\lambda)\,d\lambda \tag{3.25}$$

Proof. Using the definition of S and $W^\pm = \lim_{\varepsilon\to\pm 0} W^{(\varepsilon)}$, $W^{(\varepsilon)} = \mathbb{1} - \int R(\lambda+i\varepsilon)V\delta_\varepsilon(T-\lambda)\,d\lambda$, we obtain

$$S - \mathbb{1} = \lim_{\varepsilon\downarrow 0} W^{\pm *}(W^{(-\varepsilon)} - W^{(\varepsilon)}) \tag{3.26}$$

substituting into here the expression for $W^{(\varepsilon)}$ and taking into account the intertwinning property of $W^{\pm *}$ one finds (3.25). □

Now we proceed the proof of theorem 3.20. The properties of T_λ follow directly from lemma 3.21 and conditions (α) and (β). To obtain $S(\lambda) = \mathbb{1}_\lambda + 2\pi i T_\lambda$, we replace $W^{\pm *}$ in (3.25) by $\mathbb{1} - \lim_{\varepsilon'\downarrow 0}\int\delta_{\varepsilon'}(T-\nu)VR(\nu+i\varepsilon')\,d\nu$, take the diagonal $\varepsilon' = \varepsilon$ in passing to the limit, apply to the obtained equation Π_λ from the left and Π^*_λ, from the right and integrate out the variable ν (to justify this integration one uses conditions (α) and (β)). □

. Scattering matrix. Two-space case

Recall that the scattering operator and scattering matrix are defined as $S = W^{+*}W^-$ and $\Pi S\Pi^* = \int^\oplus S(\lambda)\,d\lambda$, where $\Pi = \{\Pi_\lambda\}$ is a unitary operator from $\hat{H}$ to a fiber direct integral $\int^\oplus \hat{H}_\lambda\,d\lambda$ associated with $\hat{H}$.

Proposition 3.23. Define the operators $\Pi^{\pm} \equiv \Pi W^{\pm *}$ acting from H to $\int^{\oplus} \hat{H}_\lambda d\lambda$. They obey $(\Pi^{\pm} f)(\lambda) = \Pi^{\pm}_\lambda f$ for $f \in H$, where $\Pi^{\pm}_\lambda = \Pi_\lambda Q(\lambda+i0)$. Moreover, $\Pi^{\pm}\Pi^{\pm *} = \mathbb{1}$.

Proof. The first part follows from statement (b) of the theorem 3.13, the second part is a trivial consequence of the isometry of $W^{\pm}$. □

Using (3.12), (2.1) and the second resolvent equation

$$R(z)J = J\hat{R}(z) - R(z)I\hat{R}(z) \quad (\text{recall: } I = HJ - J\hat{H}), \tag{3.27}$$

we transform the stationary expression for $W^{\pm}(\Delta)$ to the useful form

$$W^{\pm}(\Delta) = J\hat{E}(\Delta) - \lim_{\varepsilon \to \pm 0} \int_\Delta R(\lambda+i\varepsilon)I\delta_\varepsilon(\hat{H}-\lambda)d\lambda \tag{3.28}$$

A statement analogous to theorem 3.20 is valid also in the two space scattering theory. The correct generalization of the notion of the $T(z)$ - operator, defined in subsection 3d, is

$$T(z) = J^*I - I^*R(z)I \tag{3.29}$$

on $\hat{H}$. Obviously that for $J = \mathbb{1}$ it becomes $V - VR(z)V$.

Theorem 3.24. Let there exist Banach spaces $\hat{X}$ and $X \subset H$ (densely) such that (a) $\hat{X}$ is $\hat{H}$-smooth; (b) I extends to a bounded operator from $\hat{X}'$ to X; (c) $Q(z)$ can be extended to a family of bounded operators from X to $\hat{X}$ which has strong boundary values on $\mathbb{R}/\sigma_p(H)$. Then (i) $S(\lambda) - \hat{\mathbb{1}}_\lambda = 2\pi i \Pi^{+}_\lambda I \Pi^{*}_\lambda \equiv 2\pi i \Pi_\lambda Q(\lambda+i0) I \Pi^{*}_\lambda$ and therefore is uniformly bounded in $\lambda \in \mathbb{R}$. If, in addition, (d) $\varepsilon\hat{R}(\lambda+i\varepsilon)$ and J^*J extend to bounded operators on $\hat{X}$, the former uniformly in $\varepsilon \in \mathbb{R}$ and $\lambda \in \mathbb{R}$, then (ii) the strong limits $T_\lambda \equiv \Pi_\lambda T(\lambda+i0)\Pi^{*}_\lambda$ exist for almost all $\lambda \in \mathbb{R}$ and form a measurable uniformly bounded in $\lambda \in \mathbb{R}$ family of operators on $\hat{H}_\lambda$ (note, that $\hat{H}_\lambda$ is locally λ-independent) and $S(\lambda) = \mathbb{1}_\lambda + 2\pi i T_\lambda$.

Remarks 3.25. (α) If we replace condition (c) of the theorem by the following (c') $I^*R(z)$, $z \in \mathbb{C} / \mathbb{R}$, can be extended to a family of bounded operators on X which have strong boundary values on $\mathbb{R}/\sigma_p(H)$ and require in addition (c") J^*I is bounded from $\hat{X}'$ to $\hat{X}$, in other words if we two-space mimic the conditions of the one-space theorem 3.20, then the simple proof of the latter theorem goes over to the two-space case. The reason that we avoid conditions (c') and (c") is that they are not satisfied in the many-particle theory with J which arises naturally there. However one might manage to find an equivalent J' (i.e. $(J'-J)e^{-i\hat{H}t} \overset{s}{\to} 0$ as $|t| \to \infty$) so that (c') and (c") are obeyed with this J'. Then our objection over the use of (c') and (c") is removed.

(β) The following precedure of constructing of the spaces X and $\hat{X}$ is used in the applications. Let I be factorizable as $I = AB$, where A is a bounded operator from an auxiliary Hilbert space K to H and B is a $\hat{H}$-smooth operator from $\hat{H}$ to K. We define $\hat{X} = B^*K$ and $X = AK$. Then (b) is obviously satisfied and (a) follows from lemma 3.18.

Proof. (i) Using representation (3.28) we find

$$W^- - W^+ = 2\pi i \lim_{\varepsilon \downarrow 0} \int \delta_\varepsilon (H-\lambda) I \delta_\varepsilon (\hat{H}-\lambda) d\lambda. \tag{3.30}$$

Lemma 3.26. For any $u,v \in \hat{X}$, $<\delta_\varepsilon(\hat{H}-\lambda)u,v>$ converges in the weak-* topology of L^∞ to $<\Pi_\lambda^* \Pi_\lambda u,v>$ as $\varepsilon \downarrow 0$, i.e.

$$\int f(\lambda) <\delta_\varepsilon(\hat{H}-\lambda)u.v> d\lambda \to \int f(\lambda) <\Pi_\lambda^* \Pi_\lambda u,v> d\lambda \quad \text{on } \varepsilon \downarrow 0 \text{ for any } f \in L^1(\mathbb{R}).$$

Proof. Follows from the proof of lemma 3.21 and properties of the Poisson integral (see [SW,p 41]). □

Lemma 3.27. For any $u,v \in X$, $<\delta_\varepsilon(H-\lambda)\bar{E}_p u,v>$, where $\bar{E}_p = \mathbb{1} - E_p$, converges in the weak-* topology of $L^\infty(\mathrm{loc})$ to $<\Pi_\lambda^{\pm *} \Pi_\lambda^{\pm} u,v>$ as $\varepsilon \downarrow 0$.

Proof. Taking into account equation (3.20) and the definition of Π_λ^+ we see that we have to prove that

$$\int_\Delta f(\lambda) <\delta_\varepsilon(\hat{H}-\lambda) Q(\lambda+i\varepsilon)u, Q(\lambda+i\varepsilon)v> d\lambda \to \int_\Delta f(\lambda) <\Pi_\lambda^* \Pi_\lambda Q(\lambda+i0)u, Q(\lambda+i0)v> d\lambda$$

as $\varepsilon \downarrow 0$ for any $f \in L^1(\mathbb{R})$ and any compact Δ. Since by cond. (c) $Q(\lambda+i\varepsilon)$, is strongly continuous from X to $\hat{X}$ and since the integral on the left-hand side is bounded in virtue of cond (a) by $\int_\Delta |f(\lambda)| \, ||Q(\lambda+i\varepsilon)u||_{\hat{X}} ||Q(\lambda+i\varepsilon)v||_{\hat{X}} d\lambda$, we can replace $Q(\lambda+i\varepsilon)$ on the left-hand side by $Q(\lambda+i0)$ with an arbitrary small error. Next the same inequality shows that we can further replace $Q(\lambda+i0)u$ and $Q(\lambda+i0)v$ by λ-independent $\hat{X}$-vectors. The resulting statement which we have to prove follows from properties of the Poisson integral (see [SW, p.47]) and the proof of lemma 3.21. The approximation argument above requires some explanation. By condition (a) $H_1 \equiv \hat{H} \cap \hat{X}$ is dense in $\hat{X}$. Therefore $C(\Delta,H_1)$ is dense in $C(\Delta,\hat{X})$. Since H_1 is a pre-Hilbert space, Span$\{\phi \cdot h,\ \phi \in C^\infty(\Delta),\ h \in H_1\}$ is dense in $C(\Delta,H_1)$. Therefore, since $Q(\lambda+i0)u \in C(\Delta,\hat{X})$, it can be approximated by a finite, linear combination of functions in the form $\phi \cdot h$, $\phi \in C^\infty(\Delta)$, $h \in H_1$. Therefore it suffices to verify the desired statement for such a function replacing $Q(\lambda+i0)u$. Taking out of the inner product we arrive to the statement we are explaining. □

We show now that the weak limit in (3.30) exists and compute this limit using lemmas 3.26 and 3.27. First we need the following facts:

(α) $\delta_\varepsilon(\hat{H}-\lambda)^{1/2} u \in L^2(\mathbb{R}, \hat{H})$ and $\delta_\varepsilon(H-\lambda)^{1/2} v \in L^2(\mathbb{R}, H)$

for all $u \in \hat{H}$ and $v \in H$ by lemma 2.2,

(β) $\delta_\varepsilon(\hat{H}-\lambda)^{1/2}$ is bounded from $\hat{H}$ to $\hat{X}'$ uniformly in λ and ε, since $\|\delta_\varepsilon(\hat{H}-\lambda)^{1/2}\|^2_{\hat{H}\to\hat{X}'} \leq \|\delta_\varepsilon(\hat{H}-\lambda)\|_{\hat{X}\to\hat{X}'}$, the latter term being uniformly bounded by the definition of the $\hat{H}$-smooth spaces.

(γ) $\delta_\varepsilon(H-\lambda)^{1/2}\bar{E}_p$ is bounded from H to X' uniformly in ε and λ. Indeed, $\|\delta_\varepsilon(H-\lambda)^{1/2}\bar{E}_p\|_{H\to X'} \leq \|\delta_\varepsilon(H-\lambda)\bar{E}_p\|_{X\to X'}$. Moreover for all $x,y \in X$,

$$|<\delta_\varepsilon(H-\lambda)\bar{E}_p x,y>| = |<\delta_\varepsilon(\hat{H}-\lambda)Q(\lambda+i\varepsilon)x,\ Q(\lambda+i\varepsilon)y>|$$

by eqn (3.20). The latter term does not exceed $\|\delta_\varepsilon(\hat{H}-\lambda)\|_{\hat{X}\to\hat{X}'}\|Q(\lambda+i\varepsilon)\|^2_{X\to\hat{X}}\|x\|_X\|y\|_X$ which is uniformly bounded by (β) and condition (a);

(δ) $\int\|\Pi^*_\lambda \Pi u\|^2_{X'}d\lambda \leq C\|u\|^2$, since $\|\Pi^*_\lambda\|_{\hat{H}_\lambda\to\hat{X}'}$ is uniformly bounded, by lemma 3.21, and $\int\|\Pi_\lambda u\|^2_{\hat{H}_\lambda} d\lambda = \|u\|^2$, by the definition of Π_λ (see the beginning of subsection 3d).

(ε) $\int\|\Pi^{\pm *}_\lambda \Pi^\pm_\lambda v\|^2_{X'} d\lambda \leq C\|v\|^2$, since $\|\Pi^{\pm *}_\lambda\|_{\hat{H}_\lambda\to X'}$ is uniformly bounded, by the definition of $\Pi^\pm_\lambda$, condition (a) and lemma 3.21, and since $\int\|\Pi^\pm_\lambda v\|^2_{\hat{H}_\lambda} d\lambda \leq C\|v\|^2$ by lemma 3.27 and the equality

$$\int\|\delta_\varepsilon(H-\lambda)^{1/2}u\|^2 d\lambda = \|u\|^2. \qquad (*)$$

Now let $f_\varepsilon(\lambda) = <\delta_\varepsilon(\bar{H}-\lambda)I\delta_\varepsilon(\hat{H}-\lambda)u,v>$ and $f(\lambda) = <\Pi^{\pm *}_\lambda \Pi^\pm_\lambda I \Pi^*_\lambda \Pi_\lambda u,v>$, $\bar{H} = H\bar{E}_p$ for $u \in \hat{H}$ and $v \in X$. Facts (α)-(ε) imply that $f_\varepsilon, f \in L^1(\mathbb{R})$. We want to show that

$$\int f_\varepsilon(\lambda)d\lambda \to \int f(\lambda)d\lambda \quad \text{as} \quad \varepsilon \downarrow 0. \qquad (3.31)$$

Using facts (α)-(γ) and cond (b) we obtain that $|\int_{\mathbb{R}\setminus\Delta} f_\varepsilon(\lambda)d\lambda| \leq C<\hat{E}^{(\varepsilon)}(\mathbb{R}\setminus\Delta)u,u>^{1/2}$

Thus by lemma 2.11.

$$\int_\Delta f_\varepsilon(\lambda)d\lambda \to \int f_\varepsilon(\lambda)d\lambda \quad \text{as} \quad \Delta \to \mathbb{R}, \quad \text{uniformly in } \varepsilon \in (0,1].$$

Hence it suffices to show that

$$\int_\Delta f_\varepsilon(\lambda)d\lambda \to \int_\Delta f(\lambda)d\lambda \quad \text{as} \quad \varepsilon \downarrow 0 \text{ for any compact } \Delta. \qquad (3.32)$$

We have

$$\int_\Delta (f_\varepsilon(\lambda)-f(\lambda))d\lambda = \int_\Delta <(\delta_\varepsilon(\hat{H}-\lambda)-\Pi_\lambda^*\Pi_\lambda)u, I^*\delta_\varepsilon(\bar{H}-\lambda)v>d\lambda$$

$$+ \int_\Delta <I\Pi_\lambda^*\Pi_\lambda u, (\delta_\varepsilon(\bar{H}-\lambda)-\Pi^{\pm *}\Pi_\lambda^{\pm})v>d\lambda. \qquad (**)$$

Since $I^*\delta_\varepsilon(\bar{H}-\lambda)v \in L^2(\mathbb{R}, \hat{X})$ by facts (α) and (γ) and condition (b) and $I\Pi_\lambda^*\Pi_\lambda u \in L^2(\mathbb{R}, X)$ by facts (α) and (β) and condition (b), these vector-functions can be approximated by functions from $\mathrm{span}\{f\cdot x, f \in L^2(\Delta), x\in X\}$ and $\mathrm{span}\{f\cdot x, f\in L^2(\Delta), x\in\hat{X}\cap\hat{\mathcal{H}}\}$, respectively (remember that $\hat{X} \cap \hat{\mathcal{H}}$ is dense in $\hat{X}$ since $\hat{X}$ is $\hat{H}$-smooth). This approximation together with (**) and lemma 3.26 and 3.27 implies (3.32). The latter as was noticed above suffices for (3.31) to hold (for all $u \in \hat{\mathcal{H}}$ and $v \in X$). Since X is dense in $\mathcal{H}$ and $\int|f_\varepsilon(\lambda)|d\lambda \leq C\|v\|$ by virtue of (*), we can extend relation (3.31) to all $v \in \mathcal{H}$. Since $W^{\pm} = (\mathbb{1}-E_p)W^{\pm}$, this implies that the expression under the sign of limit in (3.30) converges weakly to $\Pi_\lambda^{\pm *}\Pi_\lambda^{\pm}I\Pi_\lambda^*\Pi_\lambda$:

$$W^- - W^+ = 2\pi i\int \Pi_\lambda^{\pm *}\Pi_\lambda^{\pm}I\Pi_\lambda^*\Pi_\lambda d\lambda.$$

Since $\Pi_\lambda\Pi^* u = (\Pi\Pi^* u)(\lambda) = u(\lambda)$ and since by proposition 3.23,

$\Pi_\lambda^{\pm}\Pi^{\pm *}u = (\Pi^{\pm}\Pi^{\pm *}u)(\lambda) = u(\lambda)$, we have furthermore

$$\Pi W^{+*}(W^- - W^+)\Pi^* u = 2\pi i\int \Pi_\lambda^+ I\Pi_\lambda^* u(\lambda)d\lambda.$$

However, the left-hand side of this equality is $\Pi(S-\mathbb{1})\Pi^*$ (see (3.8)), so the first part of the theorem is proven.

(ii) Consider the operator $T(z)$. Substituting (3.20) into (3.29) and applying Π_λ from the left we arrive at

$$\Pi_\lambda T(\lambda+i\varepsilon) = \Pi_\lambda Q(\lambda+i\varepsilon)I - i\varepsilon\Pi_\lambda(J^*J-\mathbb{1})\hat{R}(\lambda+i\varepsilon)Q(\lambda+i\varepsilon). \qquad (3.33)$$

__Lemma 3.28.__ $\varepsilon\Pi_\lambda(J^*J-\hat{\mathbb{1}})\hat{R}(\lambda+i\varepsilon)Q(\lambda+i\varepsilon) \to 0$ as $\varepsilon \to +0$ in the strong $L^2(\Delta, L(X,\hat{\mathcal{H}}_\lambda))$-sense. Here Δ is any Borel, compact subset of $\mathbb{R}$ and, remember, $\|J\| \leq 1$.

__Proof.__ Let $K \equiv J^*J - \hat{\mathbb{1}}$ and $u \in X$. Taking into account condition (d) and the fact that by lemma 3.21, $\|\Pi_\lambda\|_{\hat{X}\to\hat{\mathcal{H}}_\lambda} \leq C$ we estimate

$$\varepsilon\|\Pi_\lambda K\hat{R}(\lambda+i\varepsilon)Q(\lambda+i\varepsilon)u\|_{\hat{\mathcal{H}}_\lambda} \leq \text{const }\|Q(\lambda+i\varepsilon)u\|_{\hat{X}}, \qquad (3.34)$$

where the const depends only on the size of a subset $\Delta \subset \mathbb{R}$ inside of which λ

varies. This inequality shows first that we can replace $Q(\lambda+i\varepsilon)$ in the quantity under consideration by $Q(\lambda+i0)$. Second by the same approximation argument as in the proof of lemma 3.27 we can replace further $Q(\lambda+i0)u$ by a λ-independent $\hat{X}$-vector. Therefore to prove the lemma it suffices to show that for all $x \in \hat{X}$

$$\Pi_\lambda K\hat{R}(\lambda+i\varepsilon)x \to 0 \quad \text{in the strong } L^2(\Delta,\hat{H}_\lambda)\text{-sinse.} \tag{3.35}$$

Lemma 3.29. (3.35) holds for any $x \in \hat{X}$ and with $\Delta = \mathbb{R}$.

Proof. Using the well-known relation $\hat{R}(z) = \int_0^\infty e^{-i\hat{H}t}e^{-zt}dt$ for $\operatorname{Im} z > 0$ and $\int \|\Pi_\lambda u\|^2_{H_\lambda} d\lambda = \|u\|^2$, we find

$$\varepsilon^2 \int \|\Pi_\lambda K\hat{R}(\lambda+i\varepsilon)x\|^2 d\lambda = \varepsilon^2 \int_{-\infty}^{\infty} d\lambda \int_0^\infty \int <\Pi_\lambda K e^{-i\hat{H}t}x, \Pi_\lambda K e^{-i\hat{H}t'}x> e^{i\lambda(t-t')} e^{-\varepsilon(t+t')} dtdt$$

$$\leq \varepsilon^2 \int_0^\infty \int dtdt' e^{-\varepsilon(t+t')} |\int_0^\infty d\lambda \|\Pi_\lambda K e^{-i\hat{H}t}x\|^2_{H_\lambda}| \, |\int_0^\infty d\lambda \|\Pi_\lambda K e^{-i\hat{H}t'}x\|^2_{H_\lambda}|$$

$$= (\varepsilon \int dt e^{-\varepsilon t} \|K e^{-i\hat{H}t}x\|)^2 .$$

Furthermore since $\|J\| \leq 1$

$$\|(J^*J-\mathbf{1})e^{-i\hat{H}t}x\|^2 = \|J^*Je^{-i\hat{H}t}x\|^2 - 2\|Je^{-i\hat{H}t}x\|^2 + \|x\|^2 \leq -\|Je^{-i\hat{H}t}x\|^2 + \|x\|^2.$$

Because of (SLAI) the right-hand side of this inequality has 0 as the Abel limit as $|t| \to \infty$. □

This completes the proof of lemma 3.28. □

Lemma 3.28 together with eqn. (3.33) implies the second part of theorem 3.24. □

f. Abstract multichannel systems

Let H be a self-adjoint operator acting on a Hilbert space H. Following W. Hunziker we call the triple $\alpha = (H_\alpha, H_\alpha, J_\alpha)$, where H_α is a Hilbert space, H_α is a self-adjoint operator on H_α and J_α is bounded operator from H_α into H, a channel for H if it satisfies (AI):

$$\lim_{|t|\to\infty} \|J_\alpha e^{-iH_\alpha t} u\| = \|u\|$$

and $W^\pm_\alpha \equiv W^\pm(H, H_\alpha, J_\alpha) \equiv s - \lim e^{iHt} J_\alpha e^{-iH_\alpha t}$ exist. $W^\pm_\alpha$ are called the (out-and in-) wave operators for the channel α (or simply channel wave operators).
The system $\{\alpha\}$ of channels for H is said to be complete if and only if the

following condition is satisfied

$$\Sigma_\alpha W_\alpha^{\pm} W_\alpha^{\pm *} = 1\!\!1 . \qquad (3.36)$$

Note that there is always one <u>trivial channel</u> corresponding to the <u>point spectrum of H</u>: $p = (E_p H, HE_p, 1\!\!1_p)$. The wave operators for this channel are $W_p^{\pm} = 1\!\!1_p$ and $W_p^{\pm} W_p^{\pm *} = E_p$. In the sequel we always exclude this channel from the systems of channels for H.

It is required in the multichannel scattering theory that all channels are mutually asymptotically orthogonal (independence of the channels). It means that for any α and β, two different channels,

$$\lim_{|t|\to\infty} (J_\alpha e^{-iH_\alpha t} f, J_\beta e^{-iH_\beta t} \varphi) = 0, \quad f \in H_\alpha, \quad \varphi \in H_\beta . \qquad (3.37)$$

This implies that $R(W_\alpha^{\pm}) \perp R(W_\beta^{\pm})$ if $\alpha \neq \beta$, or, what is the same, that $W_\alpha^{\pm *} W_\beta^{\pm} = 0 \; (\alpha \neq \beta)$. The latter together with the isometry of $W_\alpha^{\pm}$ gives

$$W_\alpha^{\pm *} W_\beta^{\pm} = \delta_{\alpha,\beta} 1\!\!1_\alpha .$$

We can define an abstract multichannel scattering system also in a weaker sense, replacing all the time-limits involved by the <u>Abel-limits.</u>

The abstract multichannel scattering system is a version of the two space scattering system. Indeed, define

$$\hat{H} = \oplus H_\alpha , \; \hat{H} = \oplus \hat{H}_\alpha \quad \text{and} \quad J(\oplus f_\alpha) = \Sigma J_\alpha f_\alpha . \qquad (3.38)$$

Then $W^{\pm} = W^{\pm}(H,\hat{H},J)$ can be expressed in terms of the channel wave operators $W_\alpha^{\pm}$:

$$W^{\pm}(\oplus f_\alpha) = \Sigma W_\alpha^{\pm} f_\alpha$$

and, inversely, $W_\alpha^{\pm}$ can be recovered from $W^{\pm}$:

$$W_\alpha^{\pm} = W^{\pm} \Pi_\alpha ,$$

where Π_α is the projection from $\hat{H}$ into H_α.

(AI), for every α, and (3.37) imply that (AI) is satisfied for the operators J and $\hat{H}$, defined in (3.38). Naturally, $W_\alpha^{\pm *} W_\beta^{\pm} = \delta_{\alpha,\beta}$, written in the new language, is the isometry of $W^{\pm}$, which on the other hand is a direct consequence of (AI). Condition (3.36) of the completeness of the system of channels for H is translated into the $W^{\pm}$-terms as the $W^{\pm}$-completeness.

Below we describe two situations where (3.37) holds. Both situations are realized in the N-body scattering.

<u>Lemma 3.30.</u> Let there exist a system $\{H'_\alpha\}$ of commuting operators on H

such that $H'_\alpha J_\alpha = J_\alpha H_\alpha$. (i) If $H'_\alpha - H'_\beta$ is absolute continuous then the channels α and β are asymptotically orthogonal (i.e. (3.37) holds).

(ii) If $\mathrm{Ker}(H'_\alpha - H'_\beta) = \{0\}$, then α and β are Abel-asymptotically orthogonal, i.e.(3.37), with the t-limit replaced by the Abel-limit, holds.

The proof of this lemma follows readily from the following general result

Lemma 3.31. Let A be a self-adjoint operator (then $e^{iAt} \to 0$ as $|t| \to \infty$ in the mean weak sense: $\frac{1}{T}\int_0^T \langle u, e^{iAt} v\rangle dt \to 0$ as $T \to \infty$).

(i) If A is absolute continuous, then $\exp(iAt) \overset{w}{\to} 0$ as $|t| \to \infty$;

(ii) If $\mathrm{Ker}\, A = \{0\}$, then $\underset{|t|\to\infty}{\text{w-Abel-lim}} \exp(iAt) = 0$.

Proof. (i) is well-known. It is a consequence of the vanishing at the infinity of the Fourier transform of a L^1-function. (ii) follows from

$$\varepsilon\int_0^{\pm\infty} e^{iAt} e^{-\varepsilon t} dt = i\varepsilon(A+i\varepsilon)^{-1} \overset{w}{\to} \text{proj. on Ker}(A) \quad \text{as } \varepsilon \to \pm 0.$$

Another situation where (3.37)-Abel is guaranteed is described in the next

Lemma 3.32. Let the set of all α's be semiordered and let for any $\beta \not\subset \alpha$, $J_\beta J_\alpha$ be H_α-smooth. Then (3.37)-Abel holds.

The proof of this statement is left as an exersize to the reader.

4. Scattering Theory for N-Body Systems

In this section we present main definitions and results concerning the scattering theory for quantum many-body systems. The proofs are given as long as they help to understand the results. The technical details are considered in consequent sections. In our exposition we follow closely section 3 describing the abstract situation.

a. Hamiltonians

Consider a system of N particles in $\mathbb{R}^\nu$, $\nu \geq 3$ with masses m_i and interacting via pair potentials $V_\ell(x^\ell)$. Here ℓ labels pairs of indices and $x^\ell = x_i - x_j$ for $\ell = (ij)$. The configuration space of the system in the center-of-mass frame is defined as $R = \{x \in \mathbb{R}^{\nu N}, \Sigma m_i x_i = 0\}$ with the inner product $(x,\tilde{x}) = 2\Sigma m_i x_i \cdot \tilde{x}_i$. Denote by V^ℓ and V_ℓ the multiplication operators on $L^2(\mathbb{R}^\nu)$ and $L^2(R)$ by the functions $V_\ell(y)$ and $V_\ell(x^\ell)$, respectively.

We assume that V^ℓ are Δ-compact, i.e., compact as operators from the Sobolev space $H_2(R^\nu)$ to $L^2(R^\nu)$. Then the operator

$$H = T + V, \quad \text{where} \quad T = - \text{Laplacian on } \mathcal{H} = L^2(R) \quad \text{and} \quad V = \Sigma V_\ell ,$$

is defined on $\mathcal{D}(T)$ and is closed there. If V is real then H is self-adjoint (see supplement I, theorem SI.4). Δ-compact potentials will be called Combes potentials.

Proposition 4.1. The potentials of the class $L^p(R^\nu) + (L^\infty(R^\nu))_\varepsilon$, where $p > \max(\nu/2,2)$ and subindex ε indicates that L^∞-component can be taken arbitrarily small, are Δ-compact. The potentials of the class $L^p \cap L^q(\mathbb{R}^\nu)$, $p > \nu > q$, are Δ-smooth (and also T-smooth).

The proof of this proposition is given in supplement (examples SI.6 and SI.12).

Partitions. Let $a = \{C_i\}$ be a partition of the set $\{1,\ldots,N\}$ into nonempty, disjoint subsets C_i, called clusters. Denote by A the set of all such partitions. A is a lattice if $b \subset a$ is set for b a refinement of a: the clusters of b are subsets of the clusters of a. The smallest partition containing two partitions a and b is denoted, as usual, by $a \cup b$, i.e., $a \cup b = \sup(a,b)$. The largest partition contained in both a and b is denoted by $a \cap b$: $a \cap b = \inf(a,b)$.

The symbol $\#(a)$ will stand for the number of clusters in a. We denote by a_{max} and a_{min} the maximal and minimal elements in A, respectively, i.e. $a_{max} = \{(1,\ldots N,)\}$, it has one cluster, and $a_{min} = \{(1),\ldots,(N)\}$, it has N clusters. A pair ℓ will be identified with the decomposition on $N - 1$ clusters, one of which is the pair ℓ itself and the others are free particles. The unions and intersections appearing below of sets labeled by partitions are understood to be taken over all partitions excluding a_{max}.

With each partition a we associate the truncated Schrodinger operator H_a. It is obtained from H by neglecting the potentials linking the different clusters in the partition a:

$$H_a = H - I_a = T + V_a \quad \text{with} \quad I_a = \sum_{\ell \not\subseteq a} V_\ell, \quad V_a = \sum_{\ell \subseteq a} V_\ell. \tag{4.1}$$

Decomposed systems. Define the configuration space of a system of N particles with the centers-of-mass of subsystems $C_i \in a$ fixed

$$R^a = \{x \in R, \quad \sum_{j \in C_i} m_j x_j = 0 \ \forall \ C_i \in a\}$$

and the configuration space of the relative motion of the centers-of-mass of the clusters $C_i \in a$

$$R_a = \{x \in R, \quad x_i = x_j \quad \text{if } i \text{ and } j \text{ belong to the same } C_k \in a\} .$$

Then $R^a \perp R_a$

$$R^a \oplus R_a = R, \quad L^2(R^a) \otimes L^2(R_a) = L^2(R) . \tag{4.2}$$

Let Δ^a and Δ_a denote the self-adjoint Laplacian on $L^2(R^a)$ and $L^2(R_a)$, respectively, and let $T^a = -\Delta^a$ and $T_a = -\Delta_a$. Introduce

$$H^a = T^a + \sum_{\ell \subseteq a} V_\ell \quad \text{on} \quad L^2(R^a)$$

Then $H_a = H^a \otimes 1\!\!1_a + 1\!\!1^a \otimes T_a$ along (4.2), i.e. H^a is obtained from H_a after the removal of the center-of-mass motion of the clusters in a. Note: $H^a = H$ for $a = a_{max}$.

The set $\tau(H) \equiv \bigcup_{a \neq a_{max}} \sigma_p(H^a)$, with the agreement $\sigma_p(H^a) = \{0\}$ for $a = a_{min}$, is called the threshold set of H.

Dilation analyticity. Let $U(\rho)f(x) = \rho^{-\nu(N-1)/2} f(\rho^{-1}x)$. Then $U(\rho)TU(\rho)^{-1} = \rho^2 T$ and $V_\ell(\rho) \equiv U(\rho)V_\ell U(\rho)^{-1}$ is the multiplication operator with $V_\ell(\rho^{-1}x^\ell)$. A Combes potential V_ℓ is said to be dilation analytic in $\mathcal{O} \subset \mathbb{C}$, $\mathcal{O} \cap \mathbb{R} \neq \emptyset$ iff $V_\ell(\rho)$, considered as an operator from $H_2(R^\ell) = \mathcal{D}(T^\ell)$ to $L^2(R^\ell)$, has an analytic continuation into $\mathcal{O}$. In this case the family $H(\rho) \equiv U(\rho)HU(\rho)^{-1}$ has an analytic continuation from $\mathbb{R}$ to $\mathcal{O}$ with the common domain $\mathcal{D}(T)$. Note also that if V_ℓ is dilation analytic in $\mathcal{O}$ it is dilation analytic in the sector $A = \{z \in \mathbb{C}, |\arg z| < \alpha\}$, where $\alpha = \sup\{|\arg z|, z \in \mathcal{O}\}$.

Balslev-Combes lemma. Let H be a N-body quantum Hamiltonian with real dilation analytic potentials. Then

$$\sigma_d(H(\zeta)) \cap \mathbb{R} = \sigma_p(H) \setminus \tau(H) \quad \text{for } \operatorname{Im} \zeta \neq 0.$$

Proof. See supplement II. □

b. Channels

A many-body system is multichannel iff $\sigma_p(H^a) \neq \emptyset$ for some $a \neq a_{min}, a_{max}$. This statement will be justified in what follows. For a partition a with $\sigma_p(H^a) \neq \emptyset$ we denote by $m(a)$, $\lambda^{a,m}$ and $\psi^{a,m}$, the number of eigenvalues (counting multiplicities, $m(a)$ could be ∞), the m-th eigenvalue and the corresponding eigenfunction, respectively, (the eigenvalues are repeated in accordance with their multiplicities). For the notational convenience we set $\sigma_p(H^a) = \{0\}$ and $m(a) = 1$ for $a = a_{min} = \{(1),\ldots,(N)\}$.

We denote by lower Greek letters, $\alpha, \beta, \ldots$, the pair (a,m), where $a \in \mathcal{A}$ with $\sigma_p(H^a) \neq \emptyset$ and $1 \leq m \leq m(a)$. For $\alpha = (a,m)$ we set $a(\alpha) = a$. Thus the eigenvectors and the corresponding eigenvalues of H^a will enjoy the notations ψ^α and λ^α, respectively, with $a(\alpha) = a$. We define the channel spaces and Hamiltonians

$$\mathcal{H}_\alpha = L^2(R_a) \quad \text{and} \quad H_\alpha = \lambda^\alpha + T_a \quad (a(\alpha) = a),$$

the channel identification operators $J_\alpha : \mathcal{H}_\alpha \to \mathcal{H}$,

$$J_\alpha u = \psi^\alpha u \quad (\text{or } \psi^\alpha \otimes u) \quad \text{for } u \in \mathcal{H}_\alpha,$$

and the channel wave operators

$$W_\alpha^\pm = s - \lim_{t \to \pm\infty} e^{iHt} J_\alpha e^{-iH_\alpha t},$$

if the latter exist. Obviously

$$\|J_\alpha e^{-iH_\alpha t} u\| = \|u\| \quad \forall u \in \mathcal{H}_\alpha .$$

Following the definitions of section 3 we call the triple $(\mathcal{H}_\alpha, H_\alpha, J_\alpha)$ (or often the pair α which singles it out) a channel for H, if $W_\alpha^\pm$ exist.

Furthermore, we extend H_α to the entire $\mathcal{H}$ keeping the same notation: $H_\alpha = \lambda^\alpha + \mathbb{1}^a \otimes T_a$. Clearly $H_\alpha J_\alpha = J_\alpha H_\alpha$ (is not a commutation!) Since $H_\alpha - H_\beta$ is obviously absolutely continuous for $a(\alpha) \neq a(\beta)$ and $J_\alpha^* J_\beta = 0$ for $a(\alpha) = a(\beta)$

but $\alpha \neq \beta$ we have by lemma 3.30 that the distinct channels are asymptotically orthogonal:

$$\lim_{|t|\to\infty} \langle J_\alpha e^{-iH_\alpha t} f \,;\, J_\beta e^{-iH_\beta t} g \rangle = 0 \quad \text{for } \alpha \neq \beta \text{ and all } f \in \mathcal{H}_\alpha \text{ and } g \in \mathcal{H}_\beta .$$

Another proof of this fact comes (via lemma 3.32) from the statement that

$J_\beta^* J_\alpha$ are H_α - smooth if $\beta \not\subseteq \alpha$

We leave its proof to the interested reader.

As was shown in section 3 the properties of the channels discussed above imply the following properties of the wave operators:

$W_\alpha^\pm$ are isometric and mutually orthogonal: $W_\alpha^{\pm *} W_\beta^\pm = \delta_{\alpha\beta} \mathbb{1}_\alpha$

$W_\alpha^\pm$ are intertwinning for (H,H_α): $HW_\alpha^\pm = W_\alpha^\pm H_\alpha$

$$\oplus R(W_\alpha^\pm) \subset \mathcal{H}_{ac} , \tag{4.3}$$

where $\mathcal{H}_{a.c}$ is the absolute continuous subspace for H.

We say that the system $\{W_\alpha^\pm\}$ (or the system of the channels $\{\alpha\}$) is (asymptotically) complete iff $\oplus\, R(W_\alpha^\pm) = R(\mathbb{1} - E_p)$, i.e.

$$\Sigma W_\alpha^\pm W_\alpha^{\pm *} = \mathbb{1} - E_p . \tag{4.4}$$

Here E_p is eigenprojection on the point-spectrum subspace of H. Note that because of (4.3), relation (4.4) implies that

$$\sigma_{s.c.}(H) = \emptyset .$$

As defined above a quantum many-body system is a multichannel system in the sense of section 3 (or a realization of an abstract multichannel system in the sense of section 3). Moreover, it is equivalent to a two-space scattering system. Remember that the latter is constructed as

$$\hat{\mathcal{H}} = \oplus \mathcal{H}_\alpha , \quad \hat{H} = \oplus H_\alpha , \quad J(\oplus x_\alpha) = \Sigma J_\alpha x_\alpha \quad : \hat{\mathcal{H}} \to \mathcal{H} \tag{4.5}$$

and

$$W^\pm = s - \lim e^{iHt} J e^{-i\hat{H}t} .$$

c. Existence of wave operators

Once we identified a many-body system as a realization of the abstract multichannel system we can apply to it the results established for the latter in section 3.

Theorem 4.2. Let the potentials V_ℓ be such that H is self-adjoint and let besides $V_\ell \in L^p(\mathbb{R}^\nu)$, $p > \nu$. Then $W^\pm$ exist.

Proof. Using definition (4.5) we compute for $I = HJ - J\hat{H}$

$$I(\oplus u_\alpha) = \Sigma I_\alpha J_\alpha u_\alpha ,$$

where $I_\alpha \equiv I_{a(\alpha)} = \sum_{\ell \not\subseteq a(\alpha)} V_\ell$. Now in accordance with theorem 3.5 it suffices to show that

$$\int_1^\infty \|V_\ell \psi^\alpha e^{-iH_\alpha t} u\| \, dt < \infty \quad \text{for any} \quad \ell \not\subset a(\alpha) \quad \text{and some dense subset of } \mathcal{H}_\alpha .$$

We show this for $L^1 \cap L^2(R_a)$ (4.6)

Let $a = a(\alpha)$, $\ell \subset a$ and i and j de defined by $\ell \subset C_i \cup C_j$; for $a = \{C_i\}$. Let $T_{a,\ell}$ be the kinetic energy operator in the variable $x_{a,\ell} = x_{C_i} - x_{C_j}$ (i.e. $T_{a,\ell} = -\frac{1}{2}\Delta_{x_{a,\ell}}$, recall $x_{a,\ell} \in \mathbb{R}^\nu$). Then $H_\alpha - T_{a,\ell}$ commutes with V_ℓ and

$$\|V_\ell \psi^\alpha e^{-iH_\alpha t} u\| = \|V_\ell \psi^\alpha e^{-iT_{a,\ell} t} u\| .$$

By the Hölder inequality

$$\|V_\ell e^{-iT_{a,\ell} t} u\|_{L^2(dx_{a,\ell})} \leq \|V_\ell\|_{L^q(dx_{a,\ell})} \|e^{-iT_{a,\ell}} u\|_{L^r(dx_{a,\ell})}$$

with $q^{-1} + r^{-1} = 2^{-1}$. Now since $\ell \not\subset a$ a simple computation shows that

$$\|V_\ell\|_{L^q(dx_{a,\ell})} = \|V_\ell\|_q .$$

Now the kernel of $e^{-iT_{a,\ell} t}$ can be explicitly evaluated:

$$(2\pi it)^{-\nu/2} e^{i|x-x'|^2/2t} = (2\pi it)^{-\nu/2} e^{i|x|^2/2t} e^{i(x'x')/t} e^{i|x|^2/2t} .$$

Treating $e^{-iT_{a,\ell} t}$ as a composition of multiplication operators (by $e^{-i|x|^2/2t}$) and the Fourier transform and using the $L^p \to L^{p'}$ $(p \leq 2)$ boundedness of the Fourier transformation we get the estimate

$$\|e^{-iT_{a,\ell} t} u\|_{L^r(dx_{a,\ell})} \leq (2\pi)^{-\nu(1/r-1/2)} |t|^{-\nu(1/r-1/2)} \|u\|_{L^{r'}(dx_{a,\ell})} . \quad (4.7)$$

Putting the estimates above together and integrating along the remaining subspace $R^a \oplus R_b = R \ominus \{x_{a,\ell}\}$, $b = a \cup \ell$, we get

$$\|V_\ell \psi^\alpha e^{-iH_\alpha t} u\| \leq Ct^{-\nu(1/s-1/2)} \|V_\ell\|_q \|u\|_s , \quad \frac{1}{q} = \frac{1}{s} - \frac{1}{2}$$

taking here $s^{-1} > 2^{-1} + \nu^{-1}$ and therefore $q > \nu$, (4.6) follows. □

<u>Remark 4.3.</u> Applying more subtle estimates we can further relax the conditions on the potentials in theorem 4.2. Namely, the following estimate is true

$$\|V_\ell \psi^\alpha e^{-iH_\alpha t} u\| \leq C(u)\|(1+|\cdot|)^\alpha V_\ell\|_q |t|^{-3/q+\alpha} , \quad \alpha \geq 0, \quad 2 \leq q \leq \infty,$$

for any $u \in C_0^\infty(R_a)$. Taking here $q = 2$ and $\alpha = \frac{1}{2} - \varepsilon$ we obtain (4.6).

d. <u>Asymptotic completeness</u>

We begin with Banach spaces which are central to our treatment of the boundary values of the resolvent. We will see later that they are natural for the many-body problem. We begin with some definitions and notations. We denote by $a(b)$ a pair (b,a) of partitions such that $b \subset a$ and $\#(b) = \#(a) + 1$. The set of all such pairs can be obtained as $\{(b,a) : a = b \cup \ell$ for some $\ell \not\subset b\}$. With each pair $a(b)$ we associate the multiplication operator

$$J^\delta_{a(b)} u = (1 + |x^a_b|^2)^{\delta/2} u(x) .$$

Next we define smeared spectral projections for H_α. Let

$$\kappa = \min_{a(\beta) \subset a(\alpha)} (\lambda^\beta - \lambda^\alpha) . \tag{4.8}$$

Note, that $\kappa > 0$ since condition <u>(IE) is equivalent to</u>

$$\lambda^\beta > \lambda^\alpha \quad \text{if} \quad a(\beta) \subset a(\alpha) \quad \text{for} \quad a(\alpha) \neq a_{max} \tag{4.9}$$

Let

$$\chi \in C^\infty(\mathbb{R}) , \quad \chi(t) = 1 \ \text{if} \ t > -\frac{1}{10}\kappa \ \text{and} \ \chi(t) = 0 \ \text{if} \ t < -\frac{1}{5}\kappa \tag{4.10}$$

We define

$$\chi_\alpha(\gamma) = \chi(H_\alpha - \gamma) \quad \text{if} \quad a(\alpha) \neq a_{min} \quad \text{and}$$
$$\chi_\alpha(\gamma) = \mathbb{1} \quad \text{if} \quad a(\alpha) = a_{min} . \tag{4.11}$$

We define the Banach space $\hat{H}_\gamma = \oplus H_{\alpha,\gamma}$, where

$$H_{\alpha,\gamma} = \chi_\alpha(\gamma)^{-1}(\Sigma J^\delta_{c(a)} L^2(R_a)) \quad \text{with} \quad a = a(\alpha) \quad \text{and} \quad \delta > 1 . \tag{4.12}$$

At last, we use the shortland H_γ for $H_{\alpha,\gamma}$ with $a(\alpha) = a_{min}$ and notation

$$\kappa(\lambda) = (\lambda - \frac{1}{2}\kappa,\ \lambda + \frac{1}{16}\kappa) .$$

We prove in section 8 and 9 the following

Theorem 4.4. Under conditions (SR) (or (SR')), (QB) and (IE) of Introduction, the resolvent $R(z)$ can be represented as

$$R(z)(\mathbb{1} - E_p) = J\hat{R}(z)Q(z) ,$$

where $Q(z)$, $z \in \mathbb{C} \setminus \mathbb{R}$, is an analytic family of bounded operators from H to $\hat{H}$. Moreover, $Q(z)$ can be extended to a family of bounded operators from $L^2_\delta(R)$ (actually, from H_γ) to $\hat{H}_\gamma$ with $\delta > 1$ and $\gamma \in \kappa(\mathrm{Re}\ z)$ which has strong boundary values on $\mathbb{R} \setminus \sigma_p(H)$.

Corollary 4.5. Under conditions (SR) (or (SR')), (QB), (IE) of Introduction, the system of channel wave operators $W^\pm_\alpha$ is complete.

Proof. To apply theorem 3.13 we have only to check that

(i) $\hat{H}_\gamma \cap \hat{H}$ is dense in $\hat{H}_\gamma$ and (ii) $|\varepsilon| \int_\Delta ||\hat{R}(\lambda+i\varepsilon)x(t)||^2 d\lambda \leq M||x||_{L^2(\Delta,\hat{H}_\gamma)}$

for any bounded interval Δ (M is, of course, independent of ε). (i) is trivial: $\hat{H}_\gamma$ is dense in $\hat{H}$, (ii) results from the following, stronger result:

Lemma 4.6. $\delta_\varepsilon(\hat{H}-\lambda)$ extends to a family of bounded operators from $\hat{H}_\gamma$ to its dual $\hat{H}'_\gamma$, $\gamma \in \kappa(\lambda)$, uniformly bounded in $\varepsilon \in \mathbb{R}^+$ and $\lambda \in \mathbb{R}$ and strongly continuous in $\varepsilon \in \overline{\mathbb{R}^+}$.

Proof. It suffices to consider $\delta_\varepsilon(H_\alpha - \lambda)$ on $H_{\alpha,\gamma}$ to its dual $H'_{\alpha,\gamma}$. First we notice that $\delta_\varepsilon(H_\alpha-\lambda)(1-\chi_\alpha(\gamma))$ for $\gamma \in \kappa(\lambda)$ is a family of bounded operators on $L^2(R_a)$, norm-continuous in $\lambda \in \mathbb{R}$ and $\varepsilon \in \mathbb{R}^+$. Indeed $\mathbb{1} - \chi_\alpha(\gamma)$ is not 0 only on the subspace where $H_\alpha - \gamma \leq -1/10\kappa$. Taking into account that $\gamma \in \kappa(\lambda)$ and therefore $\gamma \leq \lambda + 1/16\kappa$ we find that on this subspace

$$H_\alpha - \lambda \leq -(\frac{1}{10} - \frac{1}{16})\kappa . \tag{4.13}$$

and hence $\delta_\varepsilon(H_\alpha - \lambda) \equiv \frac{1}{\pi}\mathrm{Im}(H_\alpha - \lambda - i\varepsilon)^{-1}$ is a nice, bounded operator.

It remains to consider the term

$$< \delta_\varepsilon(H_\alpha-\lambda)\chi_\alpha(\gamma)u,\ \chi_\alpha(\gamma)v > \ = \ \Sigma < \delta_\varepsilon(H_\alpha-\lambda)J^\delta_{d(a)}h,\ J^\delta_{c(a)}\ell > \quad \text{with } a = a(\alpha).$$

The last equality follows from the condition that $u,v \in H_{\alpha,\gamma}$ ($\chi_\alpha(\gamma)u = \Sigma J^\delta_{c(a)}h$ or $\chi_\alpha(\gamma)u = \Sigma J^\delta_{c(a)}h_c$ depending how we define $\Sigma J^\delta_{c(a)}L^2(R_a)$.

Since the both ways are equivalent and the estimates in both cases are the same we have chosen the definition which involves less notation). The desired estimate of the right-hand side of this equation follows from

Lemma 4.7. (Essentially, Kato, Iorio-O'Carroll, Combescure-Ginibre, Hagedorn). Let U and W be the multiplication operators on $L^2(R_a)$ by the functions $\varphi(x_{a,\ell})$ and $\psi(x_{a,s})$, respectively, where $\varphi,\psi \in L^p \cap L^q(\mathbb{R}^\nu)$, $p > \nu > q$, $x_{a,\ell} = x_{C_i} - x_{C_j}$, for $\ell = (ij)$ and $a = \{C_i\}$. Then $V(T_a-z)^{-1}W$ can be extended to a family of bounded operators on $L^2(R_a)$ which is analytic in $\mathbb{C}\setminus\overline{\mathbb{R}^+}$ and strongly continuous on $\mathbb{R}^+$. It is bounded in norm as

$$\|U(T_a-z)^{-1}W\| \leq \mathrm{Const}\|\varphi\|_{L^p\cap L^q}\|\psi\|_{L^p\cap L^q} .$$

Moreover, if $\ell \cap s \neq \emptyset$, then $U(T_a-z)^{-1}W$ is norm continuous as $\mathrm{Im}\, z \to \pm 0$.

Proof. For the sake of notations we take $a = a_{min}$ and omit this subindex.

We study $WR_0(z)U$ using the following representation for $R_0(z)$:

$$R_0(z) = i\int_0^{\pm\infty} e^{izt}e^{-iTt}dt, \quad z \in \mathbb{C}^\pm , \tag{4.14}$$

valid for any self-adjoint T. We consider below two cases:

(i) $\ell \cap s \neq \emptyset$. Using the equation $T = T^s + T_s$ and the fact that T_s commutes with W we find $(\alpha^{-1} + \alpha'^{-1} = 1)$

$$\|We^{-iTt}U\| \leq \|U\|_{L^2(R^s)\to L^{\alpha'}(R^s)}\|e^{-iT^st}\|_{L^{\alpha'}(R^s)\to L^{\alpha}(R^s)}\|W\|_{L^{\alpha}(R^s)\to L^2(R^s)} .$$

Writing the kernel of $\exp(-iT^st)$ as const. $t^{-\nu/2}e^{ix^2/2t}e^{ixy/t}e^{iy^2/2t}$ and using the boundedness of the Fourier transform from L^p to $L^{p'}$, $p \leq 2$, we obtain the estimate $\|e^{-iT^st}\|_{\alpha\to\alpha'} \leq$ const. $t^{-\nu(1/2-1/\alpha)}$. Taking into account this estimate and $\|W\|_{L^\alpha(R^s)\to L^2(R^s)} \leq \|\psi\|_r$ and $\|U\|_{L^2(R^s)\to L^{\alpha'}(R^s)} \leq \|\varphi\|_r$, $\alpha'^{-1} = 2^{-1} + r^{-1}$, (which follows from $\|f\cdot u\|_\alpha \leq \|f\|_r\|u\|_\beta$, $\alpha^{-1} = r^{-1} + \beta^{-1}$), we finally arrive at

$$\|We^{-itT}U\| \leq \mathrm{const}\|\psi\|_r\|\varphi\|_r t^{-\nu/r} .$$

This implies the desired properties of $WR_0(z)U$.

(ii) $\ell \cap s = \emptyset$. In this case we use the following decomposition $T = T^{\ell} + T^{s} + T_{b}$, $b = \ell \cup s$. Since T^{s}/T^{ℓ}, T_b commutes with U/W, we have

$$We^{-iTt}U = e^{-iT_b t}e^{-iT^{\ell}t}UWe^{-iT^{s}t}\,. \tag{4.15}$$

We prove now the strong continuity of $WR_0(z)U$. The estimate of its norm follows in the same way. To spare us unpleasant domain remarks, we use henceforth the extended definition of the norm:

$$\|Au\| = \infty \quad \text{if} \quad u \notin \mathcal{D}(A) \quad \text{and} \quad \|A\| = \infty \quad \text{if } A \text{ is unbounded.}$$

Using (4.14) and (4.15) we obtain

$$\|W(R_0(z') - R_0(z))Uf\| \leq \sup_{\|\varphi\|=1} \int |\langle We^{-iTt}Uf,\varphi\rangle|\,|e^{iz't} - e^{izt}|dt$$

$$\leq \{\int \|We^{-iT^{s}t}f\|^2\,|e^{iz't} - e^{izt}|^2 dt\}^{\frac{1}{2}} \sup_{\|\varphi\|=1} \{\int \|U^* e^{iT^{\ell}t}\ \|^2\, dt\}^{\frac{1}{2}}\,.$$

To complete our estimate we use Kato's inequality (see subsection 3c and supplement I, eqn (SI.7)):

$$\int \|Ae^{iTt}u\|^2\, dt/\|u\|^2 \leq \sup_{\varepsilon>0,\lambda} \|A\delta_{\varepsilon}(T-\lambda)A^*\|\,,$$

valid for any self-adjoint T and densely defined closed A, and the estimates of case (i) for $U^*R_0(z)U$ and $WR_0(z)W^*$. □

This proves lemma 4.6 and therefore corollary 4.5. □

e. Structure of the scattering matrix

We begin with a description of a fiber direct integral $\int^{\oplus}\hat{H}_{\lambda}d\lambda$ with respect to $\hat{H}$ on which the S-matrix is defined. We define $\int^{\oplus}\hat{H}_{\lambda}d\lambda$ as $L^2(\mathbb{R}, \oplus_{\alpha} L^2(\Omega_{a(\alpha)}))$, where Ω_a is a unit sphere in the dual (momentum) space R'_a (R'_a has the inner product $\langle p,q\rangle' = \Sigma(2m_i)^{-1}p_i\cdot q_i$; its norm is denoted by $|p|' = (\langle p,q\rangle')^{\frac{1}{2}}$). The unitary operator $\Pi = \int^{\oplus}\Pi_{\lambda}d\lambda$ from $\hat{H}$ to $\int^{\oplus}\hat{H}_{\lambda}d\lambda$ is defined as $\Pi_{\lambda} = \oplus\,\Pi_{\alpha}(\lambda)$, where $\Pi_{\alpha}(\lambda)$ maps $L^2(R_{a(\alpha)})$ into $L^2(\Omega_{a(\alpha)})$ in accordance with the formula

$$(\Pi_{\alpha}(\lambda)f)(\omega) = C_{\alpha}(\lambda-\lambda^{\alpha})_{+}^{m_{\alpha}}\int e^{-i\sqrt{\lambda-\lambda^{\alpha}}\,\omega\cdot x}f(x)dx\,.$$

Here $C_{\alpha} = \frac{1}{2}(2\pi)^{-\nu(N-\#(a))/2}$, $m_{\alpha} = (\nu(N-\#(a)) - 2)/4$ with $a = a(\alpha)$, $t_{+}^{\lambda} = t^{\lambda}$ for $t > 0$ and $=0$ for $t \leq 0$ and we have written the vector $p \in R'_a$ in the polar coordinates $r = |p|'$ and $\omega = p|p|'^{-1} \in \Omega_a$ as $p = r\omega$. Obviously $\Pi_{\lambda}\hat{H} = \lambda\Pi_{\lambda}$ on $\mathcal{D}(\hat{H})$.

Theorem. Under conditions (SR), (IR) and (QB), $S(\lambda) - \mathbb{1}_\lambda = 2\pi i \Pi_\lambda Q(\lambda+i0) I \Pi_\lambda^* = 2\pi i T_\lambda$, where $T_\lambda \equiv \Pi_\lambda T(\lambda+i0)\Pi_\lambda^*$, in the strong sense, with $T(z) = J^* I - I^* R(z) I$.

Proof. Assume for simplicity $|V_\ell(x)| \leq C(1+|x|)^{-\delta}$ with $\delta > 2$ for all ℓ. Go back to theorem 3.24. Condition (c) follows from the statement of theorem 4.4, condition (a) is satisfied as was demonstrated in the proof of corollary 4.5. It remains to verify that condition (b) is obeyed as well, i.e. that I (where remember $I(\oplus u_\alpha) = (HJ-J\hat{H})(\oplus u_\alpha) = \Sigma I_{a(\alpha)} J_\alpha u_\alpha)$ maps boundedly $\hat{H}'_\gamma$ into H_γ. The latter follows from the definitions of the spaces $\hat{H}_\gamma$ and H_γ and the restriction on I_a.□

f. Single-channel systems

As our basic definitions show, the channels of a many-body system are defined by the eigenvalues of its subsystems plus point 0. If no subsystem has eigenvalues, then there is only one channel. This channel is defined by the finest partition $a_{min} = \{(1),\ldots,(N)\}$ and is called the elastic channel (no bound states participate in the scattering). Such systems are called single-channel systems.

Two important classes of potentials lead to single-channel systems: weak potentials ($\|V_\ell\|_{L^p \cap L^q} \leq$ some small number, $p > \nu/2 > q$) and repulsive potentials (i.e. $\partial V_\ell / \partial |x^\ell| \leq 0$ for $x \neq 0$ [RSIV]). More generally, if each V_ℓ can be written as a sum of repulsive and $(1+|x|)^{-3-\varepsilon} L^\infty(\mathbb{R})$- small potentials, the corresponding system is single-channel ([S9]). To demonstrate this we use the results of R. Lavine [L1,2,RSIV] and a modified Iorio-O'Carroll technique [IOC,RSIV]: if $V_\ell = V_\ell^R + V_\ell^S$ in the obvious notation, then it suffices to show that the operators $UR_1(\lambda \pm i0)W$, where $|U|$ and $|W| \leq C(1+|x|)^{-3/2-\varepsilon}$ and $R_1(z) = (T+\Sigma V_\ell^R - z)^{-1}$, are bounded on $L^2(R)$. The latter can be deduced [S9] from the results of R. Lavine [L1,2] (see also [RSIV] and [Mo1]).

In the case of single-channel systems there are only two wave operators which correspond to the only channel-elastic channel. We distinguish them by the subindex 0: $W_0^\pm = s - \lim e^{iHt} e^{-iTt}$. Sometimes we introduce for convenience the coupling constants, in general complex, g_ℓ and write $H(g) = T + \Sigma g_\ell V_\ell$ and $W_0^\pm(g)$, the wave operators for $H(g)$ and T. In the single-channel case theorem 4.4 can be reformulated to a simple form. Besides we can strengthen slightly the estimates on the resolvent.

First we consider the case of weak interactions.

Theorem 4.8. Assume that, in general complex, potentials satisfy $V^\ell \in L^p \cap L^q(\mathbb{R}^\nu)$, $p > \nu/2 > q$. Then there exists a number g^0 such that for all g_ℓ with $|g_\ell| < g^0$ and all pairs ℓ and s, $|V_\ell|^{1/2}(H(g)-z)^{-1}|V_s|^{1/2}$ can be extended to an analytic in $z \in \mathbb{C} \setminus \mathbb{R}^+$ and in $g \in \{|g_\ell| < g^0 \ \forall \ \ell\}$ family of bounded operators on H strongly continuous as $z \to \mathbb{R}^+$ (uniformly in g).

Proof. We consider the second resolvent equation

$$R + RVR_0 = R_0 .$$

Multiplying it by $|V_\ell|^{\frac{1}{2}}$ from the left and by $|V_s|^{\frac{1}{2}}$ from the right we find

$$|V_\ell|^{\frac{1}{2}}R|V_s|^{\frac{1}{2}} + \sum_r |V_\ell|^{\frac{1}{2}}R|V_r|^{\frac{1}{2}}g_r V_r^{\frac{1}{2}}R_0|V_s|^{\frac{1}{2}} = |V_\ell|^{\frac{1}{2}}R_0|V_s|^{\frac{1}{2}} , \qquad (4.16)$$

where we use the notation $V_\ell^{\frac{1}{2}} = \mathrm{sign}(V_\ell)|V_\ell|^{\frac{1}{2}}$. We consider (4.16) as a matrix equation for the matrix $[|V_\ell|^{\frac{1}{2}}R|V_s|^{\frac{1}{2}}]$. By lemma 4.7, $|V_r|^{\frac{1}{2}}R_0|V_s|^{\frac{1}{2}}$ is a uniformly bounded strongly continuous up to $\overline{\mathbb{R}^+}$ family on H. Therefore for g_ℓ small enough, $\|[g_\ell V_\ell^{\frac{1}{2}}R_0|V_s|^{\frac{1}{2}}]\| < 1$ and eqn (4.16) has a unique solution with all the properties described in the theorem. □

Corollary 4.9. (Iorio-O'Carroll). Under the conditions of theorem 4.8, but with g_ℓ and V_ℓ real, $W_0^{\pm}(g)$ exist, are analytic in g for $g \in \{|g_\ell| < g^0\}$ and $W^{\pm}(g)W^{\pm}(g)^* = \mathbb{1}$. Hence $\sigma_p(H(g)) = \sigma_{s.c}(H(g)) = \emptyset$.

Now we consider the general single-channel systems.

For given pair potentials, the set of all coupling constants g, for which $H(g)$ is single-channel, is

$$\{g \mid \sigma_p(H^a(g)) = \emptyset \ \forall a, \ a \neq a_{min}, a_{max}\}$$

Let G be the interior of this set. Then $H(g_0)$ with $g_0 \in G$ remains single-channel if g_0 is slightly changed (e.g. under small perturbations). Such systems will be called strongly single-channel.

The results of [S9] imply (see the remark at the beginning of proof of thm. 7.1) that an open neighbourhood of the repulsive $L^p \cap L^q$ - potentials produces strongly single-channel systems.

Remark 4.10. We show in section 7 that a system is strongly single-channel iff it is single-channel and all of its two-particle subsystems have no quasibound states (at 0).

Theorem 4.11. Let the potentials V_ℓ, be dilation-analytic and satisfy $V_\ell \in L^p \cap L^q(\mathbb{R}^\nu)$, $p > \nu/2 > q$. Then for all pairs ℓ and s, $|V_\ell|^{\frac{1}{2}}(H(g)-z)^{-1}|V_s|^{\frac{1}{2}}$ is extendable to an analytic in $z \in \mathbb{C}\setminus\overline{\mathbb{R}^+}$ and $g \in G$ family of bounded operators on H which has strong boundary values on $\mathbb{R}^+$.

The proof of theorem 4.11 is given in section 7.

Corollary 4.12. Under the conditions of theorem 4.11 and for real g_ℓ and V_ℓ, $W_0^{\pm}(g)$ exist, are analytic in $g \in G$ and are complete: $W_0^{\pm}(g)W_0^{\pm}(g)^* = \mathbb{1} - E_p(g)$, where $E_p(g)$ is the eigenprojection on the subspace of the point spectrum of $H(g)$.

Before proceeding to the theorem on a structure of the S-matrix, we describe a fiber direct integral with respect to T. We define $\int^{\oplus} H_\lambda d\lambda$ as $L^2(\mathbb{R}^+, L^2(\Omega))$, i.e. $H_\lambda = L^2(\Omega)$. Here Ω is the unit sphere in R', the dual space to R, described at the beginning of the section. The unitary operator $\Pi = \int^{\oplus} \Pi_\lambda d\lambda$ from H to $\int^{\oplus} H_\lambda d\lambda$ is defined as

$$(\Pi_\lambda f)(\omega) = C_N \lambda^m \int e^{-i\sqrt{\lambda}\omega \cdot x} f(x)dx \ , \tag{4.17}$$

$$m = (\nu(N-1)-2)/4 \ , \quad C_N = \frac{1}{2}(2\pi)^{-\nu(N-1)/2} \ .$$

Obviously, $\Pi_\lambda T = \lambda \Pi_\lambda$ on $\mathcal{D}(T)$ and

$$\Pi_\lambda = \lambda^{-\frac{1}{2}} \Pi_1 U(\sqrt{\lambda}) \ , \tag{4.18}$$

where $U(\rho)$ is the delation group defined in subsection 4a.

Lemma 4.13. Let ℓ be a pair of indices and let M be the multiplication operator by $f(x^\ell)$, $f \in L^p \cap L^q(R^\nu)$, $p > \nu > q$, or $f \in L^2(\mathbb{R}^\nu)$. Then $\Pi_s M$ is a uniformly bounded family of operators from $L^2(\Omega)$ to $L^2(R)$, strongly continuous in s. A similar statement is true also for $M\Pi_s^*$.

Proof. Let $f \in L^p \cap L^q$, $p > \nu > q$. Then the statement follows from lemmas 3.21 and 3.18, prop. 4.1, eqn. (4.18) and strong continuity at $\rho = 1$ of $U(\rho)$ on L^p (different underlying spaces!) If $f \in L^2$, then the statement is obtained by the application of Cauchy-Schwartz inequality to $M\Pi_s^* \varphi$, $\varphi \in L^2(\Omega)$, or $\Pi_s Mu$, $u \in L^2(R)$, written explicitly as an integral. □

Recall, that A is the sector of the dilation analyticity of the potentials.

Theorem 4.14. Let H be the Hamiltonian of an N-body strongly single channel system with real, dilation analytic potentials V_ℓ such that $V_\ell(\zeta) \in L^p \cap L^q(\mathbb{R}^\nu)$, $p > \nu/2 > q$, for each $\zeta \in A$. Then the scattering matrix $S(\lambda)$ has a meromorphic continuation into the sector $\lambda \in 2A \setminus [1,\infty] \ \bigcup_{a \neq a_{max}} \sigma_p)H^a(e^{-i\alpha}))$. This continuation has poles in $A \cap \mathbb{C}^-$ only at eigenvalues of $H(e^{-i\alpha})$, i.e. at the points where the miromorphic continuation of $(u,R(z)v)$ (on dilation analytic vectors u,v) from $\mathbb{C}^+$ across $\sigma(H)$ into the second Riemann sheet has its poles. If $\alpha > \pi/2$, then the poles of $S(\lambda)$ on the negative semiaxis occur at the eigenvalues of H.

Proof. We consider $T_\lambda \equiv \Pi_\lambda V \Pi_\lambda^* - \Pi_\lambda VR(\lambda+i0)V\Pi_\lambda^*$. Let $A^\pm = \mathbb{C}^\pm \cap A$. Using (4.18), we find: $\Pi_\lambda V \Pi_\lambda^* = \lambda^{-1}\Pi_1 V(\sqrt{\lambda})\Pi_1^*$. Therefore $\Pi_\lambda V\Pi_\lambda^*$ is analytic in $\lambda \in 2A$.

Next we obtain

$$\Pi_\lambda VR(\lambda+i\varepsilon)V\Pi_\lambda^* = \lambda^{-1}\Pi_1 V(\sqrt{\lambda})R(\lambda+i\varepsilon,\sqrt{\lambda})V(\sqrt{\lambda})\Pi_1^*. \tag{4.19}$$

The r.h.s. is an $L(L^2(\Omega))$-valued function, meromorphic in $\lambda \in 2A$ as long as $\lambda + i\varepsilon \notin \sigma_{ess}(H(\sqrt{\lambda})) = \bigcup_{a\neq a_{max}} [\sigma_d(H^a(\sqrt{\lambda})) + \lambda \overline{\mathbb{R}^+}]$. To establish the convergence of (4.19) as $\varepsilon \downarrow 0$ we need

Theorem 4.15. Under the conditions of theorem 4.14 and for all ℓ and s and $\zeta \in A$, $|V_\ell(\zeta)|^{\frac{1}{2}} R(z\zeta^2,\zeta)|V_s(\zeta)|^{\frac{1}{2}}$ is an analytic in $z \notin \bigcup_{a\neq a_{max}} [\sigma_d(H^a(\zeta)) + \overline{\mathbb{R}^+}]$ family of bounded operators on H with strong boundary values on $\overline{\mathbb{R}^+}$ if $\text{Im } z \cdot \text{Im } \zeta > 0$ and on $\overline{\mathbb{R}^+} \cap [\mathbb{C} \setminus (\zeta^{-2} \bigcup_{a\neq a_{max}} \sigma_d(H^a(\zeta)) + \overline{\mathbb{R}^+})]$ if $\text{Im } z \cdot \text{Im } \zeta < 0$. Moreover, z is allowed to approach $\overline{\mathbb{R}^+}$ with angles other than $\pi/2$, e.g. $z = \lambda + i\varepsilon\zeta^{-2}$, $\varepsilon \to \pm 0$. In both cases ($z = \lambda + i\varepsilon$ and $z = \lambda + i\varepsilon\zeta^{-2}$) the convergence is uniform in ζ from any compact subset of $(\text{Re } z)^{-1}[2A \setminus [1,\infty) \bigcup_{a\neq a_{max}} \sigma_d(H^a(e^{\pm i\alpha}))]$ for $\text{Im } z \gtrless 0$.

The proof of this theorem is just a slight modification of the proof of theorem 4.11. It can be found in [S6], we omit it here.

It follows from theorem 4.15 and lemma 4.13 that (4.19) converge in the $L^2(\Omega)$-operator norm as $\varepsilon \downarrow 0$, uniformly in λ from any compact subset of $B \equiv 2A \setminus [1,\infty] \bigcup_{a\neq a_{max}} \sigma_d(H^a(e^{-i\alpha}))$ (take $z = 1 + i\varepsilon\lambda^{-1}$ and $\zeta = \sqrt{\lambda}$ in theorem 4.15). Then by the theorem on uniform convergence of analytic functions, the boundary value of (4.18) as $\varepsilon \downarrow 0$ is a meromorphic function in B, which can have poles only where $R(\lambda+i0,\sqrt{\lambda})$ does. All the poles of the latter family are in A^- and at the eigenvalues of $H(e^{-i\alpha})$. The converse is also true. Indeed, for ϕ, F, two dilation vectors, and $\lambda \in 2A$ we have $\langle\phi,R(\lambda+i0,\sqrt{\lambda})F\rangle = \langle\phi(\bar{\zeta}),R(\lambda,\sqrt{\lambda}\zeta)F(\zeta)\rangle$, where $\zeta \in A^-$ and $\phi(\zeta)$ is an analytic continuation of $U(\rho)\phi$. If $\lambda \in 2A^+$, then taking $\zeta = \lambda^{-\frac{1}{2}} \in A^-$ we see that the r.h.s. of this equality has no poles at all. If $\lambda \in 2A^-$, we can take $\zeta = \lambda^{-\frac{1}{2}}e^{-i\alpha} \in A^-$ to convince ourselves that it has poles exactly at eigenvalues of $H(e^{-i\alpha})$. □

5. Exact Parametrices

In this section we construct exact parametrix families for the family $H-z$. The rest of these lectures is devoted to the demonstration of certain properties of these families. From these properties we derive the desired estimates on the boundary values of the resolvent $R(z)$.

Definition 5.1. Let $\mathcal{H}$ be a Banach space and T, an operator on $\mathcal{H}$ with a domain $\mathcal{D}(T)$. We call a bounded operator F from $\mathcal{H}$ to $\mathcal{D}(T)$ a (right, exact) parametrix for T iff (i) F is invertible and (ii) $TF - \mathbb{1}$, raised to some power, is compact on $\mathcal{H}$.

In this section we construct parametrices for the family $H-z$ and use them to study the spectral properties of H.

We begin with some general properties of exact parametrices which brought about their use in the scattering theory. In the sequel H and G denote an operator (in general unbounded and nonself-adjoint) on a Hilbert space $\mathcal{H}$, and an open set in $\mathbb{C}$.

Theorem 5.2. Let there exist a family $F(z) : G \to L(\mathcal{H},\mathcal{D}(H))$ with a bounded inverse $F(z)^{-1} : G \to L(\mathcal{D}(H),\mathcal{H})$, such that the operator $(H-z)F(z) - \mathbb{1}$, raised to some power, is compact on $\mathcal{H}$. Then $\sigma_{ess}(H) \subset \mathbb{C} \setminus G$.

Remark. This theorem can be regarded as a generalization of Weyl's theorem. Indeed, let $H = T + V$, where V is T-compact. Taking $F(z) = (T-z)^{-1}$ for $z \in \rho(T)$ we see immediately that the conditions of theorem 5.2 are satisfied and therefore $\rho(T) \cap \sigma_{ess}(H) = \emptyset$. Reversing the role of H and T we get $\rho(H) \cap \sigma_{ess}(T) = \emptyset$. If $\sigma_d(T) = \emptyset$, then $\sigma_{ess}(H) = \sigma(T)$.

Proof. Introduce the operator $A(z) = (H-z)F(z)$. Then

$$H - z = A(z)F(z)^{-1}. \tag{5.1}$$

Therefore in virtue of the properties of $F(z)$, $\text{Ker}(H-z) = F(z)\text{Ker}\,A(z)$.

Since $A(z) - \mathbb{1}$, raised to some power, is compact, $A(z) \restriction \mathcal{H} \ominus \text{Ker}\,A(z)$ has a bounded inverse on $\mathcal{H}$. Therefore, in virtue of (5.1), so has $H - z$, restricted to $\mathcal{D}(H) \ominus F(z)\text{Ker}\,A(z)$. Since $\dim \text{Ker}\,A(z) < \infty$, the latter conclusion implies that $z \notin \sigma_{ess}(H)$. □

Lemma 5.3. Let $G \subset \mathbb{C} \setminus \sigma_{ess}(H)$ and $F(z)$ be a parametrix for $H - z$ for all $z \in G$. Let there exist a Banach space X such that $X \cap \mathcal{H}$ is dense in X

and in H and

(1) $(H-z)F(z) - \mathbb{1}$ is bounded on X and strongly continuous as $z \to \partial G$ and, raised to some power, is compact for all $z \in \overline{G}$.

Assume the following condition is satisfied:

(2) There is a unitary representation, $\rho \to U(\rho)$, of $\overline{\mathbb{R}^+}$ on H such that $U(\rho)HU(\rho)^{-1}$ and $U(\rho)A(z)U(\rho)^{-1}$ have analytic continuations $H(\zeta)$ and $A(z,\zeta)$, to a domain $\mathcal{O} \subset \mathbb{C}$, $\mathcal{O} \cap \overline{\mathbb{R}^+} \neq 0$, and $A(z,\zeta) - \mathbb{1}$ is compact for $z \in G$ and continuous in $z \in \overline{G} \setminus \sigma_{ess}(H(\zeta))$ on H and on X.

Then $0 \in \sigma(A(z)) \Longleftrightarrow z \in \sigma_p(H)$ for $z \in \overline{G} \setminus \sigma_{ess}(H(\zeta))$, $\mathrm{Im}\zeta \neq 0$.

Proof. In order to avoid lengthy expressions we assume here that $A(z) - \mathbb{1}$ is compact and norm continuous itself (and not only a power). We begin with

Lemma 5.4. Let $G,\mathcal{D} \subset \mathbb{C}$ and $K(z,\zeta)$ be a family of compact operators, jointly norm continuous on $\overline{G} \times \mathcal{D}$ and analytic in $\zeta \in \overline{\mathcal{D}} / \mathcal{D}$ as $z \in G/\overline{G}$. Let moreover $K(z,\zeta) = U(|\zeta|)K(z,e^{i \arg \zeta})U(|\zeta|)^{-1}$, where $U(\rho)$ is unitary for $\rho \in \overline{\mathbb{R}^+}$. Then for $z \in \overline{G}$ and $\nu \neq 0$

$\nu \in \sigma(K(z,\rho))$, $\rho \in \partial\mathcal{D}$, $\Longleftrightarrow \nu \in \sigma(K(z,\zeta))$, $\zeta \in \mathcal{D}$.

Proof. For $z \in G$ the statement is obtained by the standard Combes argument : $\sigma(K(z,\zeta))$ is analytic in $\zeta^{1/p}$ for some integer $p > 0$ (by the analytic perturbation theory) and is, in the same time, independent of $|\zeta|$ (by the condition of the theorem). Therefore it is ζ-independent.

Consider now $z \in \partial G$. We use

Lemma 5.5. Let $K(\lambda)$ be a continuous family of compact operators, $\lambda \in \mathcal{Q}$, a closed set, and $\nu \in \sigma(K(\lambda_0))$. Then there exists a neighbourhood V of λ_0 and $\nu(\lambda) \in \sigma(K(\lambda))$, continuous in $\lambda \in V \cap \mathcal{Q}$, for which $\nu(\lambda_0) = \nu$.

Proof. For any contour $\Gamma \subset \rho(K(\lambda_0))$ around the point ν there is a small disk, U, with the centre at λ_0, such that $\Gamma \subset \rho(K(\lambda))$ for $\lambda \in U$.

Therefore we can write $P_\lambda = (2\pi i)^{-1}\oint_\Gamma (z-K(\lambda))^{-1}dz$ for $\lambda \in U$. Since P_λ is continuous in $\lambda \in U$ and $P_{\lambda_0} \neq 0$, we have $P_\lambda \neq 0$, for $\lambda \in U$. Therefore $\sigma(K(\lambda)) \cap V \neq \emptyset$ for $\lambda \in U$ and the neighbourhood V with $\partial V = \Gamma$. □

We continue with the proof of lemma 5.4. Let $\rho \in \partial\mathcal{D}$. Assume for $z_0 \in \partial G$, $\nu \in \sigma(K(z,\rho))$. By lemma 5.5, there is $\nu(z)$, continuous in a G-neighbourhood of z_0 and such that $\nu(z_0) = \nu$ and $\nu(z) \in \sigma(K(z,\rho))$. Since $\sigma(K(z,\zeta))$ is independent of ζ as long as $K(z,\zeta)$ is analytic in ζ (see the first paragraph of the proof), $\nu(z) \in \sigma(K(z,\zeta))$. Since $K(z,\zeta)$ is norm-continuous in $z \in \overline{G}$ for $\zeta \in \mathcal{D}$, $\nu(z_0) \in \sigma(K(z_0,\zeta))$ by theorem 5.7 below.

To prove the opposite direction we repeat our arguments in the reverse order. □

Further, the proof of lemma 5.3 goes as follows. Applying lemma 5.4 to $A(z) - 1\!\!1$, we find that $0 \in \sigma(A(z)) \Leftrightarrow 0 \in \sigma(A(z,\zeta))$ for $z \in \partial G \setminus \sigma_{ess}(H(\zeta))$, $\operatorname{Im}\zeta \neq 0$. Since $A(z,\zeta) - 1\!\!1$ is norm continuous in $z \in \overline{G}$ as long as $z \notin \sigma_{ess}(H(\zeta))$ and is compact on $\mathcal{H}$ and on X for $z \in G$, it is also compact for $z \in \partial G \setminus \sigma_{ess}(H(\zeta))$. Next, we use corollary 5.9 below, stating that if a compact operator K is defined on Banach spaces X and Y and $Y \cap X$ dense in X and Y, then the spectra of K on X and Y coincide. We apply this corollary to X, $\mathcal{H}$ and $A(z,\zeta) - 1\!\!1$ to conclude that $\sigma(A(z,\zeta))$ on X and on $\mathcal{H}$ is the same. This and the definition of $A(z,\zeta)$ produce: $0 \in \sigma(A(z,\zeta)) \Leftrightarrow z \in \sigma_d(H(\zeta))$. Finally, $z \in \sigma_d(H(\zeta)) \Leftrightarrow z \in \sigma_p(H)$ for $z \cap \partial G \cap (\mathbb{C} \setminus \sigma_{ess}(H(\zeta))$ by Balslev-Combes lemma (see subsection 4a and supplement II). □

Theorem 5.6. Assume in addition to the conditions of lemma 5.3 that for some family of operators $T(z)$ with $\mathcal{D}(T(z)) \subset \mathcal{D}(H)$, $T(z)F(z)$ extends to a family of bounded operators on X, strongly continuous in $z \subset \overline{G}$. Then $T(z)(H-z)^{-1}$ extends to a family of bounded operators on X, strongly continuous in $z \in \overline{G} \setminus (G \cap \sigma_p(H))$.

Proof. The proof follows from lemma 5.3 and eqn (5.1). □

Now we will prove a few general statements owed to the proof of lemma 5.3.

Theorem 5.7. Let a sequence of compact operators K_n converge in norm to K. If $\nu_n \in \sigma(K_n)$ and $\nu_n \to \nu$ as $n \to \infty$, then $\nu \in \sigma(K)$. (In particular, the set of singular points of a norm-continuous family of compact operators is closed).

Proof. The case $\nu = 0$ is obvious. Assume $\nu \neq 0$. Let φ_n be a normalized eigenfunction of K_n corresponding to the eigenvalue ν_n, $n = 1,2,\cdots$. Since K is compact (the set of all compact operators is closed in the operator-norm topology), the sequence $\{\psi_n = K\varphi_n\}$ contains a converging subsequence $\{\psi_{n'}\}$. Then $\psi \equiv \lim \psi_{n'}$ is an eigenfunction of K with the eigenvalue ν:

$$\|(K-\nu)\psi\| \leq (\|K\| + |\nu|)\|\psi-\psi_{n'}\| + \|K\|\,\|K-K_{n'}\|\,\|\varphi_{n'}\| + \|K\|\,\|(K_{n'} - \nu_{n'})\varphi_{n'}\|$$

$$+ \|K\|\,|\nu - \nu_{n'}|\,\|\varphi_{n'}\| \to 0$$

$$\|\psi\| = \lim \|\psi_{n'}\| = \lim \|(K-K_{n'})\varphi_{n'} + \nu_{n'}\varphi_{n'}\|$$

$$= |\nu|. \qquad \square$$

<u>Theorem 5.8.</u> Let X and Y be Banach spaces with X embedded densely and continuously in Y. Let K be a compact operator on X and on Y (if i is the embedding of X into Y, then iK = Ki on X). Then the spectra of K and their multiplicities on these spaces are same. Hence, the eigenvectors of K on Y associated to non-zero eigenvalues belong also to $X \subset Y$.

<u>Proof.</u> Let $\lambda \neq 0$. First, $\mathrm{Ker}_X(K-\lambda) \subset \mathrm{Ker}_Y(K-\lambda)$. Since the embedding $X \subset Y$ is dense and continuous, the dual spaces X' and Y' obey $Y' \subset X'$. Therefore, (K' is adjoint to K)

$$\mathrm{Ker}_{X'}(K'-\lambda) \subset \mathrm{Ker}_{Y'}(K'-\lambda) .$$

Taking into account that by Riesz-Schauder theorem

$$\dim \mathrm{Ker}_X(K-\lambda) = \dim \mathrm{Ker}_{X'}(K' - \lambda)$$

and the same for Y, we obtain $\mathrm{Ker}_X(K-\lambda) = \mathrm{Ker}_Y(K-\lambda)$. $\square$

<u>Corollary 5.9.</u> Let Banach spaces X and Y be related as follows: there exists a third Banach space Z embedded densely and continuously into both X and Y (e.g. if X and Y are subspaces of some vector space with $X \cap Y$ dense in both X and Y, then we can take $Z = X \cap Y$ equipped with the norm $\|z\|_Z = \max(\|z\|_X, \|z\|_Y)$. Then the conclusion of theorem 5.8 is still true.

Now we proceed to the actual construction of the parametrics in the N-body case.

<u>Definition 5.10.</u> Let A be a finite lattice and $\{H_a, a \in A\}$ a collection of operators on H with $\mathcal{D}(H_0) \subset \mathcal{D}(H_a)$, where $H_0 = H_{a_{min}}$ (as above, a_{min} and a_{max} are the minimal and maximal elements in A). We define by introduction on $a \in A$ the following <u>$(A_a$ -) families</u> of bounded operators on H :

$$A_a(z) = (H_a - z)(H_0 - z)^{-1} \overrightarrow{\prod_{b \subsetneq a}} A_b(z)^{-1} \qquad (5.2)$$

where the arrow over the top of the product sign indicates the following order of the A^{-1}'s: if A_c^{-1} stands on the right of A_d^{-1} then $c \not\subset d$.

We set the family of bounded operator from $\mathcal{H}$ to $\mathcal{D}(H_0)$:

$$F_a(z) = (H_0 - z)^{-1} \prod_{b \subset a}^{\rightarrow} A_b(z)^{-1}. \tag{5.3}$$

(5.2) and (5.3) imply that

$$(H_a - z) F_a(z) = A_a(z) . \tag{5.4}$$

The obvious properties of A_a and F_a are listed for the reference convenience to the following two lemmas:

<u>Lemma 5.11</u>. For any $a \in A$ and all $z \in \bigcap_{b \subset a} \rho(H_b)$, the operator $F_a(z)$ is bounded from $\mathcal{H}$ to $\mathcal{D}(H_0)$ and has the bounded inverse (from $\mathcal{D}(H_0)$ to $\mathcal{H}$). Both operators are analytic in $z \in \bigcap_{b \subset a} \rho(H_b)$.

<u>Lemma 5.12.</u> For any $a \in A$ and all $z \in \bigcap_{b \subset a} \rho(H_b)$, the operator $A_a(z)$ is bounded on $\mathcal{H}$ and is analytic in $z \in \bigcap_{b \subset a} \rho(H_b)$. It has the bounded inverse for $z \in \bigcap_{b \subset a} \rho(H_b)$ and the following statements are equivalent.

1. $0 \in \sigma(A_a(z))$ and $\phi \in \mathrm{Ker}\, A_a(z)$

2. $z \in \sigma_d(H_a)$ and $F(z)\phi \in \mathrm{Ker}(H-z)$.

In case when the H_a's are constructed as (4.1) the operators $F_a(z)$ and $A_a(z)$ have an additional structure $(H_0 = T)$:

<u>Lemma 5.13</u>. The operators $F_a(z)$ and $A_a(z) - 1\!\!1$ are finite, linear combinations of monomials of the form

$$R_0 \Pi [V_\ell R_b], \ell, b \subset a, \quad \text{and} \quad \Pi [V_\ell R_b], b \subset a, \cup \ell = a, \tag{5.5}$$

respectively. Here $R_0(z) = (T-z)^{-1}$ and $R_b(z) = (H_b - z)^{-1}$

The statement can easily be derived by induction (see appendix I). Note here only that since V_ℓ have T-bound 0, they are H_b-bounded as well (see supplement I for the definitions). Therefore monomials of form (5.5) are bounded and analytic in $z \in \cap\rho(H_b)$.

Lemma 5.14. For z with $\mathrm{dist}(z,\sigma(T))$ sufficiently large, $A_a(z) - 1\!\!1$ is a norm convergent series of monomials,

$$\prod_{\cup\ell=a} [V_\ell (T-z)^{-1}].$$

Proof. The statement follows from lemma 5.13 and the fact that for $\mathrm{dist}(z,\sigma(T))$ large enough the following series are norm convergent

$$R_b(z) = (T-z)^{-1} \sum_{n=0}^{\infty} [\sum_{\ell \subset b} V_\ell (T-z)^{-1}]^n. \tag{5.6}$$

Indeed, $\|A(T-z)^{-1}\| \to 0$ as $\mathrm{dist}(z,\sigma(T)) \to \infty$ for any T-bounded operator A with the T-bound 0 (see supplement I, lemmas I.3).

Lemma 5.15. If V_ℓ are Combes potentials, then $A_a(z) - 1\!\!1$ is compact on H for each $z \in \cap\rho(H_a)$.

Proof. In virtue of lemma 5.11 it suffices to show that the monomials $\Pi[V_\ell R_b]$ with $\cup\ell = a_{max}$, are compact. First we introduce cut-off functions

$$X \in C^\infty(\mathbb{R}^\nu),\ X(y) = 0 \ \text{ for } \ |y| \leq 1 \ \text{ and } \ X(y) = 1 \ \text{ for } \ |y| \geq 2.$$

We define the multiplication operators X_n^ℓ on $L^2(R^\ell)$ by the functions $X(n^{-1}x^\ell)$. Clearly $X_n^\ell \to 0$, strongly, as $n \to \infty$. Therefore, since $V^\ell(T^\ell-z)^{-1}$ is compact, $X_n^\ell V^\ell(T^\ell-z)^{-1} \to 0(n \to 0)$. Moreover, we compute $[T^\ell, X_n^\ell] = n^{-2}F_n^\ell + n^{-1}G_n^\ell$. $\forall\ell$, where μ_ℓ is the reduced mass for the pair ℓ, F_n^ℓ is the multiplication operator by the function $(2\mu_\ell)^{-1}(\Delta X)(n^{-1}x^\ell)$ and the i-th component of G_n^ℓ is the multiplication operator by $\mu_\ell^{-1}(\nabla_i X)(n^{-1}x^\ell)$. Therefore $[T^\ell, X_n^\ell](T^\ell-z)^{-1} \to 0$ as $n \to \infty$. We define $X_\ell^{(n)} = X_n^\ell \times 1\!\!1_\ell$ and $\bar{X}_\ell^{(n)} = 1\!\!1 - X_\ell^{(n)}$. These operators have the properties:

(i) $X_\ell^{(n)} V_\ell R_b(z) \to 0 \quad (n \to \infty)$,

(ii) $[T, X_\ell^{(n)}] R_b(z) \to 0 \quad (n \to \infty)$,

(iii) $\Pi\bar{X}_\ell^{(n)}$ with $\cup\ell = a_{max}$ belongs to $C_0^\infty(R)$.

We claim that $\Pi\bar{X}_\ell^{(n)}\Pi[V_\ell R_b]$ converge in norm to $\Pi[V_\ell R_b]$ as $n \to \infty$. Then since the former operators are compact for $\cup\ell = a_{max}$ in virtue of (iii), the compactness of the latter will follow from the theorem on closedness of the set of compact operator in the uniform topology. The convergence is proved in two steps. First we note that because of (i), $\Pi[V_\ell \bar{X}_\ell^{(n)} R_b]$ converge in norm to $\Pi[V_\ell R_b]$ as

$n \to \infty$. On the second step we commute the $\bar{X}_\ell^{(n)}$'s in the former operator to the left to the position in front of the first R_b on the left. Because of the equation

$$[\bar{X}_\ell^n, R_b] = R_b [X_\ell^n, T] R_b$$

and property (ii), the terms containing at least one commutator $[\bar{X}_\ell^n, R_b]$ vanish in norm as $n \to \infty$. Therefore the difference between the operators $\Pi [V_\ell \bar{X}_\ell^n R_b]$ and $\Pi \bar{X}_\ell^n \Pi [V_\ell R_b]$ goes in norm to zero as $n \to \infty$. This completes the proof. □

<u>Corollary 5.16</u>. If V_ℓ are Combes potentials, then $\sigma_{ess}(H) \subset \cup \sigma(H_a)$.

<u>Degression.</u> It is easy to prove that $L(z) \equiv A(z) - \mathbb{1}$ is compact on $L^2_\delta(R)$ for any $\delta \geq 0$. Namely, the following statement is true.

<u>Lemma 5.17.</u> If V_ℓ are Combes potentials, then

$J^{-\delta} L(z) = L(z) J^{-\delta} + K(z)$ for any $\delta > 0$,

where $K(z)$ is compact on $L^2(R)$. Hence $L(z)$ is compact on $L^2_\delta(R)$, $\delta \geq 0$. Here <u>J^δ is the multiplication operator by $(1+|x|^2)^{-\delta/2}$.</u>

<u>Proof.</u> Pull $J^{-\delta}$ through $L(z)$ by "little bits" $J^{-\delta/n}$, where n is sufficiently large, using the commutator identities

$$[J^{-\alpha}, R_a(z)] = R_a(z) [H_a, J^{-\alpha}] R_a(z)$$

and that $|\nabla J^{-\alpha}| \leq C J^{1-\alpha}$ and $|\Delta J^{-\alpha}| \leq C J^{2-\alpha}$. □

6. Quasibound States and the Finiteness of the Discrete Spectrum

In this section we relate the quasibound states defined in the introduction with solutions of certain homogeneous equations associated with Fredholm equations used in the scattering theory. This connection determines the role played by the quasibound states in the scattering theory. As a by product our analysis implies the finiteness of the number of isolated eigenvalues.

The connection mentioned above is especially simple in the two-particle case: $H = \Delta + V$ has a quasibound state at 0

$$\Longleftrightarrow (-\Delta+V)\psi = 0 \quad \text{for} \quad \psi = (\Delta)^{-1}L^p(\mathbb{R}^\nu), \quad p < \frac{\nu}{2}, \quad \psi \notin L^2 \tag{6.1}$$

$$\Longleftrightarrow \varphi + V(-\Delta)^{-1}\varphi = 0 \quad \text{for} \quad \varphi \in L^p(\mathbb{R}^\nu), \quad p < \frac{\nu}{2}, \quad \Delta^{-1}\varphi \notin L^2.$$

The Hardy-Littlewood-Sobolev potential theorem (or the generalized Young inequality) and Hölder inequality imply that $V(\Delta)^{-1}$ is compact on $L^p(R)$, $p < \nu/2$, provided $V \in L^{\nu/2}(\mathbb{R}^\nu)$. In the same way one shows that $V(-\Delta+a)^{-1}$ is norm continuous as $a \downarrow 0$. The latter implies due to an abstract result at the end of this section that H has only a finite number of isolated eigenvalues.

Equivalence (6.1) is based on the equation $(H-\lambda)R_0(\lambda) = \mathbb{1} + VR_0(\lambda)$ which in the many-body case, should be replaced by the basic parametrix equation

$$(H-\lambda)F(\lambda) = \mathbb{1} + L(\lambda).$$

Since the thresholds in the many-body case are embedded into the continuous spectrum we use the dilation-analyticity to move the latter out of the way.

So let $H(\zeta)$ be the dilation-analytic family associated with the Hamiltonian H and let $A(z,\zeta)$ be the A-family for $H(\zeta)$ as defined in section 5. We also define $L(z,\zeta) = A(z,\zeta) - \mathbb{1}$ and $R(z,\zeta) = (H(\zeta)-z)^{-1}$.

In order to fix ideas we set in what follows $\operatorname{Im}\zeta < 0$. In this section we relax implicit conditions (QB) and (IE), we replace them by (here T stands for "threshold")

(T) No subsystem has bound or quasibound states at its thresholds.

On the other hand, in order to keep the exposition as simple as possible we strengthen the explicit condition: we replace (SR) by

$$|V_\ell(y,s)| \leq C(s)(1+|y|)^{-\delta} \quad \text{for some} \quad \delta > 2. \tag{6.2}$$

Let F be the Fourier transform. The image of the Fourier transform on $L^p \cap L^2(R')$ will be denoted by $F(L^p \cap L^2)(R)$. Here, recall, R' is the space dual to R (i.e. the momentum space).

The main result of this section is the following theorems.

Theorem 6.1. Let the dilation-analytic potentials satisfy (6.2) and condition (T) be obeyed. Then $L(z,\zeta)$ defines an analytic in $(z,\zeta) \in \mathbb{C}^+ \times \overline{A^-}$ family of compact operators on $F[L^2 \cap L^p](R)$, $p > \nu/\nu-2$, norm continuous in $\overline{\mathbb{C}^+}$ for $\zeta \in A^-$. Here, remember, $A^- = A \cap \mathbb{C}^-$ with A, the dilation analyticity sector.

Theorem 6.2. Let the conditions of theorem 6.1 be satisfied. Let λ^γ be an n-cluster threshold of H. (i) If $\nu(n-1) \lesssim 4$, then the following two statements are equivalent:

(α) H has a bound or/and quasibound state at λ^γ.

(β) $-1 \in \sigma(L(\lambda^\gamma,\zeta))$, $\operatorname{Im} \zeta < 0$, on $F[L^2 \cap L^p](R)$, $p > \frac{\nu}{\nu-2}$.

(ii) If $\nu(n-1) > 4$, then the following two statements are equivalent :

(α') H has a bound state at λ^γ.

(β') $-1 \in \sigma(L(\lambda^\gamma,\zeta))$, $\operatorname{Im} \zeta < 0$, on $F[L^2 \cap L^p](R)$, $p > \frac{\nu}{\nu-2}$.

The connection between solutions of the equations $\varphi + L(\lambda^\gamma,\zeta)\varphi = 0$ and $(H(\zeta)-\lambda^\gamma)\varphi = 0$ comes from the relation

$$(H(\zeta)-\lambda)F(\lambda,\zeta) = \mathbb{1} + L(\lambda,\zeta) \,, \qquad (6.3)$$

where $F(\lambda,\zeta)$ is the F-family for $H(\zeta)$ as defined in sect.5 (eqn.(5.3)).

Lemma 6.3. Let H be a many-body Schrödinger operator with real dilation analytic potentials and let $H(\zeta)$ be the dilation analytic family associated with H. Then

$$\sigma_p(H(\zeta)) \cap \tau(H) \subset \sigma_p(H)$$

If, in addition, H satisfies the conditions of theorem 6.1, then

$$\sigma_p(H(\zeta)) \cap \tau(H) = \sigma_p(H) \cap \tau(H)$$

Remark. One would wish to prove that $\sigma_p(H) \cap \tau(H) \subset \sigma_p(H(\zeta))$ with no **extra** conditions on H.

Proof. Show first that $\sigma_p(H(\zeta)) \cap \tau(H) \subset \sigma_p(H)$. Let, on the contrary, there is $\lambda \in \sigma_p(H(\zeta)) \cap \tau(H)$ such that $\lambda \notin \sigma_p(H)$. Then by the spectral theorem for self-adjoint operators $\lim_{z\to\lambda} (z-\lambda)\langle u,R(z)v\rangle = 0$ for all $u,v \in \mathcal{H}$.

Note that $\langle u,R(z)v\rangle = \langle u(\bar\zeta),R(z,\zeta)v(\zeta)\rangle$ for all dilation analytic vectors (recall that $f(\zeta)$ is an analytic continuation of $U(\sigma)f$, see paragraph (IV) of the proof of the Balslev-Combes theorem in supplement II). Hence using the density of dilation analytic vectors and the standard continuity argument, we obtain $\lim(z-\lambda)\langle u,R(z,\zeta)v\rangle = 0$ for all $u,v \in H$. But for the eigenfunction $\psi(\zeta)$ of $H(\zeta)$ corresponding to λ we have

$$\langle u,R(z,\zeta)\psi(\zeta)\rangle = (\lambda-z)^{-1}\langle u,\psi(\zeta)\rangle$$

and therefore,

$$\lim (\lambda-z)\langle u,R(z,\zeta)\psi(\zeta)\rangle = \langle u,\psi(\zeta)\rangle$$

which is different from 0 for at least some u. Thus we arrive at the contradiction. This implies that $\lambda \subset \sigma_p(H)$ and therefore $\sigma_p(H(\zeta)) \cap \tau(H) \subset \sigma_p(H)$.

Since we do not use the statement $\sigma_p(H) \cap \tau(H) \subset \sigma_p(H(\zeta))$ in this book, we only sketch its proof. Let $\lambda \in \tau(H)$ and $\lambda \not\in \sigma_p(H(\zeta))$ and show that $\lambda \not\in \sigma_p(H)$. It is shown in the course of the proof of theorem 6.1 that $\langle u,R(z,\zeta)v\rangle$ has a limit as z finds to λ from $\mathbb{C}^+$ for $u,v \in L^2_\delta(R)$ with δ sufficiently large, provided H has no quasibound states at λ. If H does have a quasibound state at λ then one can show that $|\langle u,R(z,\zeta)v\rangle| \leq C|\lambda-z|^{-1/2}$ for u and v the same as above. Thus $\lim_{z\to\lambda} (\lambda-z)\langle u,R(z)v\rangle = \lim_{z\to\lambda} (\lambda-z)\langle u(-\zeta),R(z,\zeta)v(\zeta)\rangle = 0$

for dilation analytic u and v such that $u(\zeta),v(\zeta) \in L^2_\delta(R)$ with sufficiently large δ. So $\lambda \not\in \sigma_p(H)$. □

Theorem 6.1. is a special case of cor.6.8, and theorem 6.2 follows from lemmas 6.3, 6.12, 6.13 below.

In the sequel, $H_a(\zeta)$ and $H^a(\zeta)$ are dilation-analytic families associated with H_a and H^a, respectively, $R_a(z,\zeta) = (H_a(\zeta)-z)^{-1}$ and $R^a(z,\zeta) = (H^a(\zeta)-z)^{-1}$, γ is a given threshold of H and the following terminology is frequently used: $R_b(z,\zeta)$ will be called a <u>nonsingular</u> <u>(singular)</u> resolvent iff $a(\gamma) \not\subseteq b(a(\gamma) \subseteq b)$. This definition is justified by the fact that in virtue of the Balslev-Combes theorem (see subsection 4a and supplement II) $\lambda^\gamma \in \rho(H^b(\zeta))$ in the nonsingular case and $\lambda^\gamma \in \sigma(H^b(\zeta))$ in the singular one.

V_ℓ will be called an <u>integrating potential</u> iff $\ell \not\subset a(\gamma)$. This name becomes clear after lemma 6.4 below. <u>J^δ_ℓ stands for the multiplication operator by the function $(1+|x^\ell|^2)^{-\delta/2}$.</u>

Let $X^a_p = F(L^p \cap L^2)(R^a)$ with the usual norm $\|f\|_{X^a_p} = \|Ff\|_p + \|f\|_2$.

By lemma 5.3, $L(z,\zeta)$ is a finite linear combination of operators of the form $\Pi[V_\ell R_b]$, $b \subset a$, $\cup\ell = a$. Those of these operators which actually occur in this combination will be called <u>admissable graphs</u>.

The result we want to prove first is

Proposition 6.4. For any $\ell \not\subseteq a(\gamma)$ and $\delta > 2$, the operators $J_\ell^\delta R^a(z,\zeta)$ are bounded on X_p^a, $p > \nu/\nu-2$, analytic in $(z,\zeta) \in \mathbb{C}^+ \times A^-$ and norm continuous in a $\overline{\mathbb{C}^+}$ -neighbourhood of λ^γ.

Proof. We prove the proposition by induction on partitions a. It is obvious for the trivial partition $a = \{(1)\cdots(N)\}$. We assume it to be true for all $b \subset a$ and demonstrate it for a. Below the super index a is omitted (it should be present whenever there is no other super index). We begin with

Lemma 6.5. For any $\ell \not\subseteq a(\gamma)$ and $b \subset a$, the operators $J_\ell^\delta R_b(z,\zeta)$ with $\delta > 2$ are bounded on X_p, $p > \nu/\nu-2$, and norm continuous in z in a $\overline{\mathbb{C}^+}$ -neighbourhood of λ^γ.

Proof. If $\ell \subseteq b$ then the proof goes as follows. Let F_b be the Fourier transform in R_b'. Then $F_b^{-1} J_\ell^\delta R_b(z,\zeta)u = \int_{R_b'}^{\oplus} J_\ell^\delta R^b(z-p^2)u(p)\,dp$, where $\int_{R_b'}^{\oplus} u(p)\,dp = F_b^{-1}u$. This equation together with the induction assumption about

If the partition ℓ is not a refinement of b, then we use one more step to reduce the problem to the case considered above. Using twice the second resolvent equation we find (the arguments are omitted)

$$R_b = R_d - R_d I_d R_d + R_d I_d R_b I_d R_d \tag{6.4}$$

where $d = a(\gamma)$. Since $J_\ell^\delta V_\ell(\zeta)$ is bounded and $I_d = \sum_{\substack{\ell \not\subseteq d \\ \ell \subseteq b}} V_\ell$, to complete the proof it suffices to demonstrate the proposition only for $b = d$.

To prove the proposition for $b = d$, we split R_b using $\mathbb{1} = P_\gamma(\zeta) + \bar{P}_\gamma(\zeta)$, where $P_\gamma(\zeta)$ is (eigenprojection for $H^{a(\gamma)}(\zeta)$ corresponding to $\lambda^\gamma) \otimes \mathbb{1}_{a(\gamma)}$. Since the reduced resolvent $R_b(z,\zeta)\bar{P}_\gamma(\zeta)$ is analytic at $z = \lambda^\gamma$, it has the required properties. The required properties for the other piece $R_b(z,\zeta)P_\gamma(\zeta) = (\lambda^\gamma + \zeta^2 T_b - z)^{-1} P_\gamma(\zeta)$, follow from the next lemma below. □

Lemma 6.6. Let f be the convolution operator with $f \in L^{\nu/\nu-2} \cap L^{2-\varepsilon}(\mathbb{R}^\nu)$ and $(Au)(k) = |k|^2 u(k)$. Then $f(A-w)^{-1}$ is norm continuous and uniformly bounded from $L^p(\mathbb{R}^\nu)$ to $L^2 \cap L^p(\mathbb{R}^\nu)$, $p > \nu/\nu-2$, in any compact subset from the sector $|\arg w - \pi| \leq \pi/2$.

Proof. First we apply the Hölder inequality to $(f(A-w)^{-1}u)(k) = \int \frac{f(k-q)u(q)}{q^2-w} dq$.

This gives $\| \int \frac{f(\cdot-q)u(q)}{|q|^2-w} dq \|_r \leq C \|u\|_p \| \int \frac{|f(\cdot-q)|^{p'}}{|q|^{2p'}} dq \|_{\frac{r}{p'}}^{p'}$, where $\frac{1}{p} + \frac{1}{p'} = 1$.

Now we apply the generalized Young inequality (see[RSII]) to $\int \frac{|f(k-q)|^{p'}}{|q|^{2p'}} dq$;

$$\| \int \frac{|f(\cdot-q)|^{p'}}{|q|^{2p'}} dq \|_{\frac{r}{p}}^{\frac{1}{p}} \leq \| \, |q|^{-2} \|_{\beta p', w} \|f\|_{\alpha p'}, \text{ with } \frac{1}{\alpha p'} + \frac{1}{\beta p'} = \frac{1}{p'} + \frac{1}{r} \text{ and}$$

$2p'\beta = \nu$. Here $\| \, |q|^{-2} \|_{\frac{\nu}{2}, w} < \infty$ is the weak - $L^{\frac{\nu}{2}}$ norm of $|q|^{-2}$ (see[RSII]).

Collecting all these estimates we get

$$\| \int \frac{f(\cdot-q)u(q)}{|q|^2} dq \|_r \leq C \|f\|_s \|u\|_p$$

with $\frac{1}{s} = 1 - \frac{2}{\nu} + \frac{1}{r} - \frac{1}{p}$ and $p > \frac{\nu}{\nu-2}$. Now taking r to be either p or 2 we obtain the desired boundedness estimate. The norm continuity is proved in a similar way. □

Lemma 6.7. An admissable graph is bounded on X_p, $p > \nu/\nu-2$ and norm continuous in z in a $\overline{\mathbb{C}^+}$-neighbourhood of λ^γ.

Proof. It is easy to show (see appendix I where the structure of $L(z)$ is analyzed in detail) that the admissable graphs have the property:

(*) There is at least one integrating potential on the left from each singular resolvent such that no other singular resolvent stands between them.

In the rest of the proof we omit the arguments in the operator-functions.

Let G be an admissable graph. Write each singular resolvent. R_b, in G as $p^{-\delta} J_p^\delta R_b$, where p labels the closest integrating potential on the left of R_b (so $p \not\subseteq a(\gamma)$). Next we move $J_p^{-\delta}$ to the left till it meets the mentioned potential. This move is possible since by property (*) between R_c and this potential there are no other singular resolvents. We pull J_p^δ through a nonsingular resolvent using the commutation relations

$$[J_p^{-\delta}, R_b] = e^{-2\zeta} R_b [T, J_p^{-\delta}] R_b$$

and

$$-[T, J_p^{-\delta}] = 2(\nabla J_p^{-\delta})\nabla + \Delta J_p^{-\delta}$$

and observing that $\nabla J_p^{-\delta}$ and $\Delta J_p^{-\delta}$ grow as $|x|^{\delta-1}$ and $|x|^{\max(0,\delta-2)}$ at the infinity. The multiplication operators $\nabla J_p^{-\delta}$ (and $\Delta J_p^{-\delta}$ if $\delta > 2$) should be also moved to the left. Finally either $J_p^{-\delta}$ ($\nabla J_p^{-\delta}$ or $\Delta J_p^{-\delta}$) reaches V_p and gets bounded by the latter or falls in the commutation with T. Since $J_p^{-\delta} V_p(\zeta)$ are bounded and $J_p^{\delta} R_b(z,\zeta)$ with $p \not\subseteq a(\gamma)$ and $\delta > 2$ are bounded and norm continuous by lemma 6.5, we conclude that G is bounded and norm continuous. □

Since $L(z,\zeta)$ and $J_\ell^{\delta} F(z,\zeta)$ with $\ell \not\subseteq a(\gamma)$ is a linear combination of admissable graphs, lemma 6.7 implies

Corollary 6.8. The operators $L(z,\zeta)$ are bounded on X_p, $p > \nu/\nu-2$, analytic in $\mathbb{C}^+ \times A^-$ and norm continuous in a $\overline{\mathbb{C}^+}$-neighbourhood of λ^{γ}. Hence they are also compact up to $z = \lambda^{\gamma}$.

Corollary 6.9. $J_\ell^{\delta} F(z,\zeta)$, $\ell \not\subseteq a(\gamma)$, defines an analytic family from $\mathbb{C}^+ \times A^-$ to $L(X_p)$, $p > \nu/\nu-2$, norm continuous in a $\overline{\mathbb{C}^+}$-vicinity of λ^{γ}.

In the same way one proves the following result needed below. We introduce the space related to quasibound states at the threshold λ^{γ}:

$$QB(\gamma,\zeta,p) = \psi^{\gamma}(\zeta) \otimes (\Delta_{a(\gamma)})^{-1} F[L^p \cap L^2](R_{a(\gamma)})] + (-\Delta+1)^{-1} F[L^p \cap L^2](R),$$

where, recall, $\psi^{\gamma}(\zeta)$ is the eigenfunction of $H^{a(\gamma)}(\zeta)$ corresponding to the eigenvalue λ^{γ}.

Lemma 6.10. Let $p > \nu/\nu-2$. Then $F(\lambda^{\gamma},\zeta) f \in QB(\gamma,\zeta,p)$ for $f \in X_p$ and $F(\lambda^{\gamma},\zeta)^{-1} \varphi \in X_p$ for $\varphi \in QB(\gamma,\zeta,p)$:

Finally, to apply the Fredholm alternative we need

Proposition 6.11. $-1 \notin \sigma(L(z,\zeta))$ for $z \in \mathbb{C}^+ \cup \mathbb{R}$ (remember: $\operatorname{Im}\zeta < 0$) and $a \neq a_{\max}$ (remember that a is the omitted superindex).

Proof. For $z \in \overline{\mathbb{C}^+} \setminus \{\text{thresholds of } H\}$, $-1 \in \sigma(L(z,\zeta))$ implies, due to eqn (6.3) and the fact that z is in $\cap\, \rho(H_b(\zeta))$, that $z \in \sigma_d(H(\zeta))$. This, by the Balslev-Combes theorem (see supplement II), is impossible. Hence

$$-1 \notin \sigma(L(z,\zeta)) \tag{6.5}$$

for $z \in \overline{\mathbb{C}^+} \setminus \{\text{thresholds of } H\}$. For the threshold set of H we use

Lemma 6.12. The following two statements are equivalent

(α) $-1 \in \sigma(L(\lambda^\gamma,\zeta))$

(β) $(H(\zeta)-\lambda^\gamma)\varphi = 0$ has a nontrivial (weak) solution in the space $QB(\gamma,\zeta,p)$, $p > \nu/\nu-2$.

Proof. The statement follows from eqn (6.3), the invertability of $F(z,\zeta)$, corollary 6.8 and lemma 6.10. □

Thus the two-cluster thresholds are taken care of by condition (T) formulated at the beginning of this section: $\varphi + L(\lambda^\gamma,\zeta)\varphi = 0$, $\varphi \neq 0$ would imply the existence of either a bound state at $\lambda^\gamma(\varphi \in \mathcal{D}(\Delta))$ or a quasibound state at $\lambda^\gamma(\varphi \notin \mathcal{D}(\Delta))$. Hence it remains to demonstrate (6.5) for n-cluster thresholds with $n \geq 3$. The latter is done in

Lemma 6.13. Let λ^γ be an n-cluster threshold of H with $n \geq 3$. Then $-1 \in \sigma(L(\lambda^\gamma,\zeta)) \Rightarrow \lambda^\gamma \in \sigma_p(H)$.

Proof. By corollary 5.9, lemma 6.11 and eqn (6.3) $-1 \in \sigma(L(\lambda^\gamma,\zeta))$ implies that $H(\zeta)\varphi = \lambda^\gamma\varphi$ has a nontrivial solution in $\bigcap_{p>\nu/\nu-2} QB(\gamma,\zeta,p)$.

Show now that $QB(\gamma,\zeta,p) \subset \mathcal{D}(T)$ for $p > \frac{2\nu(n-1)}{\nu(n-1)-4}$ $(\nu(n-1) > 4)$, which implies that $\bigcap_{p>\frac{\nu}{\nu-2}} QB(\gamma,\zeta,p) \subset \mathcal{D}(T)$. Let $\alpha \in C_0^\infty(\mathbb{R})$ and $\alpha(s) = 1$ for $|s| < 1$. By the Hölder inequality $\alpha(\Delta_{a(\gamma)})(\Delta_{a(\gamma)})^{-1}F(L^p)(R_{a(\gamma)}) \subset L^2(R_{a(\gamma)})$ and therefore $\alpha(\Delta_{a(\gamma)})(\Delta_{a(\gamma)})^{-1}F(L^p \cap L^2)(R_{a(\gamma)})) \subset \mathcal{D}(\Delta_{a(\gamma)})$ for $p > \frac{2\nu(n-1)}{\nu(n-1)-4}$. On the other hand, $(1-\alpha(\Delta_{a(\gamma)}))(\Delta_{a(\gamma)})^{-1}$ maps $L^2(R_{a(\gamma)})$ into $\mathcal{D}(\Delta_{a(\gamma)})$. Hence $(\Delta_{a(\gamma)})^{-1}F[L^p \cap L^2](R_{a(\gamma)})] \subset \mathcal{D}(\Delta_{a(\gamma)})$ and therefore $QB(\gamma,\zeta,p) \subset \mathcal{D}(T)$ for $p > \frac{2\nu(n-1)}{\nu(n-1)-4}$. Thus, $-1 \in \sigma(L(\lambda^\gamma,\zeta))$ implies $\lambda^\gamma \in \sigma_p(H(\zeta))$, $\mathrm{Im}\,\zeta < 0$. Hence, by the first part of lemma 6.3, $\lambda^\gamma \in \sigma_p(H)$. □

In virtue of lemma 6.13 and condition (QB), relation (6.5) holds also for any-cluster thresholds. This completes the proof of proposition 6.10. □

Corollaries 6.8 and 6.9, proposition 6.10 and eqn (6.3) imply the statement of proposition 6.4 for the partition a. This completes the induction proof of this proposition. □

Corollary 6.8 for $a = a_{max}$ $(=\{(1,\cdots,N)\})$ is exactly theorem 6.1 (remember that a is the omitted superindex). This proves theerem 6.1. □

Lemmas 6.3, 6.12 and 6.13 imply theorem 6.2. □

The proof of theorem 6.1 goes through also for non-dilation-analytic potential and the lowest threshold of H. In this case the statement takes the form

Theorem 6.14. Let $V_\ell \in L^{\nu/2}(\mathbb{R}^\nu)$ and no subsystem have bound or quasibound states at $\mu = \inf \sigma_{ess}(H)$. Then $L(\lambda)$ defines an analytic in $\lambda < \mu$ family of compact operators on $F(L^p)(R)$, $p > \nu/\nu-2$, norm continuous as $\lambda \uparrow \mu$.

As a by-product we obtain

Theorem 6.15. Let $V_\ell \in L^{\nu/2}(\mathbb{R}^\nu)$ and no subsystem has a quasibound state at its two-cluster threshold if the latter coincides with $\mu = \inf \sigma_{ess}(H)$. Then H has only a finite number of isolated eigenvalues.

The proof of this theorem follows readily from theorem 6.14, the equation $(H-\lambda)F(\lambda) = \mathbb{1} + L(\lambda)$ and the following abstract result ([S5])

Theore, 6.16. Let H be a self-adjoint operator on $\mathcal{H}$, $\mu \in \mathbb{R}$ and for each $\lambda < \mu$ there exists a parametrix, $F(\lambda)$, for $H - \lambda$. Assume, in addition that there are Banach spaces X with $X \cap \mathcal{H}$ dense in X and in $\mathcal{H}$ and $Y \subset X'$ such that

(i) $F(\lambda)$ is bounded from X to Y and is strongly continuous as $\lambda \uparrow \mu$ and $\mathrm{Ker}\, F(\mu) = \{0\}$.

(ii) $(H-\lambda)F(\lambda) - \mathbb{1}$ is compact on X and norm continuous as $\lambda \uparrow \mu$.

Then the spectrum of H in $(-\infty,\mu)$ consists of, at most, a finite number of eigenvalues of finite multiplicities.

Remark 6.17. Conditions (i) and (ii) might be considerably relaxed. For instance, it suffices that the compactness and norm continuity hold only for some power of $(H-\lambda)F(\lambda) - \mathbb{1}$.

Proof. First of all notice that $(-\infty,\mu) \cap \sigma_{ess}(H) = \emptyset$ by theorem 5.2. Let W be a negative H-compact operator on $\mathcal{H}$ which is also compact from Y to X. Consider a family

$H(g) = H + gW.$

We associate with this family the two-parameter family

$$A(\lambda,g) = A(\lambda) + gWF(\lambda) = (H(g)-\lambda)F(\lambda), \tag{6.6}$$

where $A(\lambda) = (H-\lambda)F(\lambda)$. Assume $0 \in \sigma(A(\mu))$. Let $\nu(g)$ be an eigenvalue of $A(\mu,g)$ such that $\nu(0) = 0$ and $\varphi(g)$, a corresponding eigenvector. By the Rellich-Kato analytic perturbation [K2] theory $\nu(g)$ and $\varphi(g)$ are analytic in a broken power of g near $g = 0$. (In the next paragraph we show that they are

We show now that $\frac{d\nu}{dg}(0)$ exists and $\neq 0$. Denote $\nu_1 = g^{-1}\nu(g)$, $\varphi(0) = \varphi \in \text{Ker } A(\mu)$, $\Delta\varphi = \varphi(g) - \varphi$ and $\varphi_1 = g^{-1}\Delta\varphi$. Applying $f \in \text{Ker } A(\mu)^*$ to the equation,

$$A(\mu)\varphi_1 + WF(\mu)\varphi = \nu_1\varphi + (\nu_1 - WF(\mu))\Delta\varphi$$

and sending g to 0 we get that if either $\langle\varphi,f\rangle \neq 0$ or $\langle WF(\mu)\varphi,f\rangle \neq 0$ then $\nu'(0)$ exists and satisfies the equation

$$\nu'(0)\langle\varphi,f\rangle = \langle WF(\mu)\varphi,f\rangle .$$

Observe that $A(\mu)^*F(\mu) = F(\mu)^*A(\mu)$ implies $F(\mu)\varphi \in \text{Ker } A(\mu)^*$. So we can pick $f = F(\mu)\varphi$ to obtain

$$\nu'(0) = \langle WF(\mu)\varphi,F(\mu)\varphi\rangle / \langle\varphi,F(\mu)\varphi\rangle .$$

So $\nu'(0) \neq 0$ as we claimed and we conclude that

$$0 \in \sigma(A(\mu)) \Rightarrow 0 \notin \sigma(A(\mu,g)) \quad \text{for } g \neq 0 \text{ sufficiently small.} \tag{6.7}$$

Since W is H-compact, $(-\infty,\mu) \cap \sigma_{ess}(H(g)) = \emptyset$ for all g. If H has infinite number of eigenvalues $<\mu$ (which might accumulate only to μ) then so has $H(g)$ for all positive g, by the comparison theorem, since $W < 0$ and therefore $H(g) < H$. Hence by eqn (6.6) and corollary 5.9, $0 \in \sigma(A(\lambda_n(g),g))$ for all $g \geq 0$, where $\lambda_n(g) \uparrow \mu$ $(n \to \infty)$ are eigenvalues of $H(g)$. By the closedness of the set of singular points theorem (theorem 5.7 with $\nu_n = \nu = 0$), $0 \in \sigma(A(\mu,g))$ for all $g \geq 0$ which contradicts conclusion (6.7) reached before. □

7. Boundary Values of the Resolvent. General Single-Channel Case.

In this section we derive the estimates of the resolvent of a single-channel H, which were discussed in Section 4f (see Theorem 4.11). The treatment below contains already all the main ideas of the proof in the general case given in the next section. At the same time it is much simples than the mentioned proof and is rather transparent.

Below we introduce the parameter $g = (g_\ell)$, coupling constant, into the formula replacing everywhere V_ℓ by $g_\ell V_\ell$ for all ℓ's. It suffices for us to keep g real. However, we do not use this restriction and therefore omit mentioning it explicitely. We define the domains for the coupling constants: $G^a = \mathrm{Int}\ \{g: \sigma_p(H^b(g)) = \emptyset \text{ for all } b \subsetneq a\}$. Remember that a system, described by $H = T + \Sigma V_\ell$, is called strongly single-channel if the systems described by $H(g) = T + \Sigma g_\ell V_\ell$ are single-channel for all g in a vicinity of $(g_\ell = 1)$ (i.e. if it remains single-channel under small perturbations). It follows from eqn (7.4), lemmas 7.10, 7.11, 7.12 below that a single-channel, short-range system is strongly single-channel iff H^ℓ has no quasibound state at 0 for any pair ℓ (i.e. for any ℓ, $H^\ell\psi = 0$ has no nontrivial non - L^2 - solution in $(\Delta^\ell)^{-1}|V^\ell|^{\frac{1}{2}}L^2(R^\ell)$).

<u>Theorem 7.1.</u> Let the potentials be dilation analytic and $V_\ell \in L^p \cap L^q(\mathbb{R}^\nu)$, $p > \frac{\nu}{2} > q$. Then for each $a \in A$ and any pair $\ell, s \subset a$, the family $|V_\ell|^{\frac{1}{2}}(H^a(g)-z)^{-1}|V_s|^{\frac{1}{2}}$ is uniformly bounded on H^a, analytic in $z \in \mathbb{C}/\sigma(T)$ and $g \in G^a$ and strongly continuous as $z \to \sigma(T)$, uniformly in $g \in G^a$. Here $H^a = L^2(R^a)$.

<u>Proof.</u> We conduct the proof by induction on $a \in A$. For $a = a_{min}$ we do not have H^a. Let the statement be true for all $b, b \subset a$, and prove it for a. In the sequel we <u>suppress the supperindex a.</u> We will prove in fact a slightly weaker statement which covers only those strongly single-channel systems for which there is a path in the complex g-space joining $g = (1)$ with $g = (0)$ and such that each of its points corresponds to a strongly single-channel Hamiltonian $H(g) = T + \Sigma g_\ell V_\ell$. To prove the general case one can use the method of section 8 which instead of the analytic continuation in g uses approximation of $(T^b-z)R^b(z)$ by operators on H^b with smooth fast vanishing integral kernels. We use the method below because of a certain elegance it possesses. In addition we conjecture that for any given $L^p \cap L^q$-potential the set $\{g: H(g)$ is strongly single channel$\}$ is simply complex connected and therefore contains a neighbourhood of $g = (0)$.

To prove the desired statement we employ the resolvent equation derived in section 5:

$$R(z,g)A(z,g) = F(z,g),$$

where $R(z,g) = (H(g)-z)^{-1}$ and $A(z,g)$ and $F(z,g)$ are A- and F-families, respectively, for the operator $H(g)$ as defined in section 5 (eqns.(5.2) and (5.3)).

Consider the operator $L(z,g) = A(z,g) - \mathbb{1}$. Writing

$L(z,g) = \sum_\ell |V_\ell|^{1/2} L_\ell(z,g)$ and defining $L_{\ell s} = L_\ell |V_s|^{1/2}$

and

$F_{\ell s} = V_\ell^{1/2} F |V_s|^{1/2}$ we obtain

$$V_\ell^{1/2} R |V_s|^{1/2} + \sum_f V_\ell^{1/2} R |V_f|^{1/2} L_{fs} = F_{\ell s} \tag{7.1}$$

Proposition 7.2. The operators $L_{\ell s}(z,g)$ and $F_{\ell s}(z,g)$ are bounded on $\mathcal{H}$, analytic in $z \in \mathbb{C}/\sigma(T)$ and $g \in G$ and strongly continuous as $z \to \sigma(T)$ uniformly in $g \in G$.

Proof. The operators $L_{\ell s}$ and $F_{\ell s}$ are linear combinations of

$$\overset{\rightarrow}{\prod_{i=1}^{k}} [V_{f_i}^{1/2} R_{b_i} |V_{f_{i+1}}|^{1/2}], \quad b_i \neq a_{max}, \tag{7.2}$$

$f_1 = \ell$, $f_{k+1} = s$, with the condition $Uf_i = a_{max}$ in the case of $L_{\ell s}$.

We transform (7.2) so that each factor satisfies $f_i, f_{i+1} \subseteq b_i$ if $b_i \neq a_{min}$. To this end we use the equations $R_b = F_b - R_b L_b$ and $R_b = F_b' - L_b' R_b$, where $F_b'(z) = F_b(\bar{z})^*$ and $L_b'(z) = L_b(\bar{z})^*$, to surround each R_b in (6.2) with V_ℓ, next on its left, and $|V_s|^{1/2}$, next on its right, satisfying $\ell, s \subset b$.

Lemma 7.3. The operator $|V_\ell|^{1/2} R_b(z,g) |V_s|^{1/2}$, $\ell, s \subset b \subset a$, considered on $\mathcal{H}$, are bounded, analytic in $z \in \mathbb{C}/\sigma(T)$ and $g \in G^b$ and strongly continuous as $z \to \sigma(T)$ uniformly in $g \in G^b$.

Proof. Let $S_b = \mathbb{1}^b \otimes s_b$, where s_b is a unitary operator from $L^2(R_b)$ to the direct integral $\int^\oplus \mathcal{H}_\lambda d\lambda$ with respect to T_b: $(s_b T_b^a u)(\lambda) = (s_b u)(\lambda)$. Then ($g$ is omitted)

$$S_b |V|^{1/2} R_b(z) |V_s|^{1/2} u = |V_\ell|^{1/2} R^b(z-\lambda) |V_s|^{1/2} (S_b u)(\lambda) .$$

This equation together with the induction statement about $|V_\ell^b| R^b(z) |V_s^b|^{1/2}$ implies the lemma. Here V_ℓ^b is the restriction of V_ℓ to $L^2(R^b)$. □

If $b_i = a_{min}$, then we use

Lemma 7.4. (Kato, Iorio-O'Corroll, Combescure-Ginibre, Hagedorn). Let U and W be the multiplication operators by functions $\phi(x^\ell)$ and $\psi(x^s)$, where ϕ, $\psi \in L^p \cap L^q(\mathbb{R}^\nu)$, $p > \nu > q$, and ℓ and s are arbitrary pairs of indices. The family $W(-\Delta - z)^{-1}U$ is bounded on H, analytic in $z \in \mathbb{C}/\mathbb{R}^+$, has strong boundary values on $\mathbb{R}^+$ and is bounded in norm by

$$||W(-\Delta - z)^{-1}U|| \leq \text{const}||\phi||_{L^p \cap L^q}||\psi||_{L^p \cap L^q} . \tag{7.3}$$

Moreover, if $\ell \cap s \neq \emptyset$, then the family is norm continuous as $\text{Im } z \to \pm 0$.

Proof. This is a special case of lemma 4.7 (caution: the index a is used in lemmas 4.7 and 7.4 in different qualities; we should set $a = a_{min}$ in lemma 4.7). □

Lemmas 7.3 and 7.4, the remark about $L_{\ell s}$ and $F_{\ell s}$ made in the paragraph preceding lemma 7.3 imply the statement of proposition 7.2. □

Proposition 7.5. The matrix $[L_{\ell s}(z,g)]^3$, $g \in G$, is compact for all $z \in \mathbb{C} \setminus \mathbb{R}$ up to the real axis.

Proof. Since the matrix is analytic in $g \in G$, it suffices to prove the proposition for a neighbourhood of $g = (0)$. It follows from lemma 7.4 that the series (5.6) with $V_\ell \to g_\ell V_\ell$ converges in the norm on $\Sigma|V_\ell|^{\frac{1}{2}}H$ for all $z \in \mathbb{C}/\mathbb{R}$ up to the real axis, as long as g is confined to a neighbourhood of 0. Substituting such series for R_b's in (7.2) we conclude that $L_{\ell s}(z,g)$ for g in a neighbourhood, V, of zero is a norm convergent series of terms of the form

$$g^k \prod_{Uf_i = a} [V_{f_i}^{\frac{1}{2}} R_0 |V_{f_{i+1}}|^{\frac{1}{2}}].$$

Lemma 7.6. Let U_ℓ and W_ℓ be the multiplication operators by functions $\phi_\ell(x_\ell)$ and $\psi_\ell(x^\ell)$, respectively, where $\phi_\ell, \psi_\ell \in L^p \cap L^q(\mathbb{R}^\nu)$, $p > \nu > q$. Then a product of three operator-functions $\mathbb{C} \setminus \mathbb{R}^+ \to L(H)$ of the form $\Pi[W_{\ell_i}(-\Delta - z)^{-1}U_{\ell_{i+1}}]$, $U\ell_i = a$, (called the a-connected graphs) has norm-continuous boundary values on $\overline{\mathbb{R}^+}$. These boundary values are compact. (Remember that all operators and spaces involved possess the superindex a which is omitted from their notations. e.g. R stands here for R^a).

Proof. We begin with

Lemma 7.7. Graphs $\Pi[W_{\ell_i}(T-z)^{-1}U_{\ell_{i+1}}]$ are norm continuous in $\phi_\ell, \psi_\ell \in L^p \cap L^q(\mathbb{R}^\nu)$, $p > \nu > q$, uniformly in $z \in \mathbb{C} \setminus \mathbb{R}^+$

Proof. The statement follows from basic estimate (7.3). □

Lemma 7.8. The product of three a-connected graphs with $\phi_\ell, \psi_\ell \in C_0^\infty(\mathbb{R}^\nu)$ is norm continuous on H as $\operatorname{Im} z \to \pm 0$.

We will prove this lemma at the end of the section. Now we deduce the proof of lemma 7.6 from lemmas 7.7 and 7.8. Indeed, since C_0^∞ is dense in L^p, there exist sequences $\phi_\ell^{(n)}$ and $\psi_\ell^{(n)}$ from C_0^∞ converging in $L^p \cap L^q(\mathbb{R}^\nu)$, $p > \nu > q$, to ϕ_ℓ and ψ_ℓ, respectively. Given a graph G we construct the new graphs, $G^{(n)}(z)$, by replacing in $G(z)$ all U_ℓ and W_ℓ by the operators $U_\ell^{(n)}$ and $W_\ell^{(n)}$ of multiplication by $\phi_\ell^{(n)}(x^\ell)$ and $\psi_\ell^{(n)}(x^\ell)$, respectively. By lemma 7.7, $G^{(n)}(z) \to G(z)$ in norm, uniformly in $z \in \mathbb{C} \setminus \mathbb{R}^+$, as $n \to \infty$. Now consider the product of three a-connected graphs and the norm approximation to this product constructed as above. By lemma 7.8, this approximation is norm continuous as $\operatorname{Im} z \to \pm 0$. Hence the product itself is norm continuous as $\operatorname{Im} \to \pm 0$. This completes the proof of lemma 7.6. □

Now we return to the proof of proposition 7.5. As was noticed in the beginning of this proof, the matrix $[L_{\ell s}(z,g)]^3$ for $g \in V$ is a norm convergent series of terms each of which is, in virtue of lemma 7.6, a compact operator on $\oplus H$ for all $z \in \mathbb{C} \setminus \mathbb{R}$ up to the real axis. By the theorem on the closedness of the set of compact operators in the uniform topology, $[L_{\ell s}(z,g)]^3$ is compact as well (for $g \in V$). □

Proposition 7.9. $-1 \notin \sigma[L_{\ell s}(z,g)]$.

Proof. (The parameter g is omitted henceforth). First we note that $-1 \in \sigma[L_{\ell s}(z)] \Leftrightarrow -1 \in \sigma(L(z))$ on $\Sigma |V_\ell|^{1/2} H$, by the construction. Furthermore, lemma 5.3 implies that

$$-1 \in \sigma(L(z)) \Leftrightarrow z \in \sigma_p(H) \quad \text{for } z \in \mathbb{C} \setminus \{0\} . \tag{7.4}$$

Hence it remains to demonstrate that this correspondence holds also for $z = 0$, the only threshold of H (remember that $\sigma_p(H^b) = \emptyset$, $b \neq a_{min}$, since the system is single channel). We begin with

Lemma 7.10. The following two statements are equivalent:

(α) $-1 \in \sigma(L_{\ell s}(z)]$

(β) $(H-z)\phi = 0$ has a nontrivial weak solution in $R_0(z)\Sigma|V_\ell|^{1/2}H$.

Proof. If $f + [L_{\ell s}(z)]f = 0$, $f = \oplus f_\ell \in \oplus H$, then $\chi = \Sigma|V_\ell|^{1/2} f_\ell$ satisfies $A(z)\chi = 0$ and therefore, in virtue of $A(z) = (H-z)F(z)$, $\phi = F(z)\chi$ obeys formally (in the weak sense) $(H-z)\phi = 0$. Since $F(z)$ has a bounded inverse we can go backward as well. □

Below we consider the cases $\#(a) = N-1$ (i.e. a can be identified with a pair ℓ) and $\nu(N-\#(a)) > 4$ separately (they overlap at $\#(a) = N-1$, $\nu \geq 5$). (We reintroduce the superindex $a = \ell$ in the first case and keep it out in the second).

Lemma 7.11. Let $V_\ell \in L^p \cap L^q(\mathbb{R}^\nu)$, $p > \frac{\nu}{2} > q$. Then for all internal points of $G^\ell \equiv \{g\colon \sigma_p(H^\ell(g)) = \emptyset\}$ the equation $H^\ell(g)\psi = 0$ has no non trivial solution in $(T^\ell)^{-1}|V^\ell|^{\frac{1}{2}}L^2(R^\ell)$.

Proof. Let, on the contrary $H^\ell(g)\psi = 0$ with $\psi \in (T^\ell)^{-1}V^\ell L^2(R^\ell)$ and $\psi \neq 0$. Then $-g^{-1} \in \sigma(V^\ell(T^\ell)^{-1})$ on $|V_\ell|^{\frac{1}{2}}L^2(R^\ell)$. By the perturbation theory (we use here the fact that $V^\ell(T^\ell-\lambda)^{-1}$ is norm continuous as $\lambda \uparrow 0$) for any sufficiently small $\lambda < 0$ there exists g' such that $-g'^{-1} \in \sigma(V^\ell(T^\ell-\lambda)^{-1})$ and $g' \to g$ as $\lambda \to 0$. The latter implies that $\lambda \in \sigma_d(H(g'))$ for g' as close to g as we wish. However, this is impossible since g is an internal point of G^ℓ. □

Lemma 7.12. $H\psi = 0$, $\psi \in (T)^{-1}\sum|V_\ell|\mathcal{H}$ implies that either $\psi = 0$ or $0 \in \sigma_p(H)$ for all a with $\nu(N-\#(a)) > 4$.

Proof. Let $\psi \in T^{-1}|V_\ell|^{\frac{1}{2}}\mathcal{H}$, then $F(0)^{-1}\psi \in \Sigma|V_\ell|^{\frac{1}{2}}\mathcal{H}$. If moreover, ψ is a solution to $H\psi = 0$, then $\phi = F(0)^{-1}\psi$ is a solution to $\psi + L(0)\phi = 0$. Proposition 7.5, corollaries 6.8 and 5.9 and theorem 6.13 imply that if $\phi \in \sum|V_\ell|^{\frac{1}{2}}\mathcal{H}$, $\phi \neq 0$, obeys the latter equation then $0 \in \sigma_p(H)$. □

Proposition 7.9 is proven. □

Proposition 7.2, 7.5 and 7.9 imply the statement of theorem 7.1 for H^a. This completes out inductive proof. □

Proof of Lemma 7.8. Lemma 7.8 is a rather simple special case of lemma 8.15 proved in section 8 and appendix III by using a complex distortion technique. Here we outline another proof lemma 7.8 (actually, of a stronger statement). The details of the latter proof can be found in [S2] where it is conducted under conditions satisfied in our case.

We set $\Delta_x^\nu(h)f(x,y) = |h|^{-\nu}(f(x+h,y)-f(x,y)$ if $0 < \nu \leq 1$ and $\Delta_x^\nu(h)f(x,y) = f(x,y)$ if $\nu = 0$.

Lemma 7.13. Let $G(z)$ be a product of three a-connected graphs with $\phi_\ell, \psi_\ell \in C_0^\infty(\mathbb{R}^\nu)$. Then the Fourier transform $G(p,q,z)$, of its kernel satisfies the estimate

$$|\Delta_{p,q}^\nu(h)\Delta_z^\mu(w)G(p,q,z)| \leq \text{const.}\ (1+|p-q|)^{-r}, \quad r \in \mathbb{R}^+.$$

Here p and q are two sets of independent variables in the space dual to R^a (i.e. in the corresponding momentum space).

Sketch of the proof. The expression for the kernel of $G(z)$ in the momentum representation (i.e. the Fourier transform of the kernel) can be easily computed, since the kernels of U_ℓ, W_ℓ and $(T^a-z)^{-1}$ in this representation are known. It has the following form

$$G(p,q,z) = \int \frac{\phi(p,q,k)d^m k}{\prod_1^s [P_i(p,q,k)-z)}, \tag{7.5}$$

where $\phi(p,q,k) \in C^\infty$ comes from the potential part (U_ℓ and W_ℓ) of $G(z)$ and $P_i(p,q,k)$ is the symbol of T^a expressed in the variables p,q,k, using an i-dependent linear transformation. The estimate of the decay of $G(p,q,z)$ at infinity can be easily obtained if we note that those of the P_i's with large enough p_k or q_k (say $p_k^2 > 10$ Rez+1) are not singular in the sense that $P_i - \text{Re}z \geq \delta > 0$. An estimation of the decay of $G(p,q,z)$ in such a p_k or q_k is a rather simple but, unfortunately, boring and longsome exercise. Since moreover, the precise form of the estimating function is not important (what is important is its L^1-property) we omit here the derivation of the infinity-decay estimate.

To obtain the smoothness estimates for those variables p_j and q_j which stay in the bounded region of $\mathbb{R}^\nu$ and the smoothness estimates in z, we join those P_i, which contain variables (counting also the k-variables) from the vicinity of infinity specified above, to ϕ.

The resulting integral is of the form

$$J(u,z) = \int \frac{\phi(k)d^m k}{\prod_1^s [(x,R^i x)-z]}, \quad x = (k,u), \quad k \in \mathbb{R}^{\nu m}, \quad u \in \mathbb{R}^{\nu n}, \tag{7.6}$$

where u varies in a compact region of $\mathbb{R}^{\nu n}$, $\phi \in C_0^\infty(\mathbb{R}^{\nu m})$ and R^i are real, non-negative, $(m+n) \times (m+n)$-matrices, R^i act on the space $\mathbb{R}^{\nu(m+n)}$, of which vectors are written as $p = (p_1 \ldots p_{m+n})$, $p_i \in \mathbb{R}^\nu$, according to the equation

$$(Rp)_i = \sum_{j=1}^{m+n} (R)_{ij} p_j .$$

To obtain the desired estimates on (7.6) we use, first the Feymann identity,

$$\prod_{i=1}^{s} A_i^{-1} = \int_{[0,1]^s} \left(\sum_i \alpha_i A_i \right)^{-s} \delta(1-\Sigma\alpha_i) d^s\alpha \quad ,$$

to transform the product of s polynomials (of the second degree) in the denominator into one polynomial (also of the second degree) but taken to the s-th power. Then we integrate by parts in k. □

Remark 7.14. To study the operator-matrix $[L_{\ell s}(z)]$ on $\oplus H$ is the same as to study the single operator $L(z)$ on $\sum_\ell |V_\ell|^{\frac{1}{2}} H$. The former way spares us of some extra explanations which would accompany the proof otherwise, while the latter one admits a generalization to the multichannel case.

8. Boundary Values of the Resolvent. The General Case

In this section we prove estimates on the boundary values of the resolvent $R(z)$ on the real axis only under conditions (SR), (QB), (IE), formulated in the introduction. To this end we study the paramatrix families constructed in section 5. The main ideas are the same as in section 7, except that instead of the analyticity in the coupling constant (which does not hold in the multichannel case) we use the norm continuity in the potentials and non-singular parts of the resolvents of the subsystems ($(T-z)R_a(z)$ in the singlechannel case) to prove the compactness result. However, the spaces and especially the derivations of the strong continuity of $L(z)$ and $F(z)$ become more involved, the latter owing much to the difficult combinatorial problems.

a. Truncated Hamiltonians

Before proceeding to the main theme we elaborate on the intermediate configuration spaces and truncated Hamiltonians introduced in section 4. First we note that $R^b \subset R^a$ and $R_a \subset R_b$ for $b \subset a$, so we can define $R^a_b = R^a \ominus R^b = R_b \ominus R_a$. Then $R^b_c \oplus R^a_c = R^a_c$ and $L^2(R^a_c) = L^2(R^b_c) \otimes L^2(R^a_b)$ for $c \subseteq b \subseteq a$. Note that $R_b = R^a_b$ if $a = a_{max}$ and $R^a = R^a_b$ if $b = a_{min}$. Recall that $\mathcal{H}^a = L^2(R^a)$.

The operator T^a_b on $L^2(R^a_b)$ is defined as the self-adjoint extension of $-\Delta^a_b$, where Δ^a_b is the Laplacian on R^a_b. Then $T^a = T^a_b$ for $b = a_{min}$ and $T_b = T^a_b$ for $a = a_{max}$. Moreover

$$T^a_c = T^b_c \times 1\!\!1^a_b + 1\!\!1^b_c \times T^a_b .$$

We define $H^a_b = T^a + \sum_{\ell \subseteq b} V_\ell$ on $\mathcal{H}^a$ and use the shorthand $H^a = H^a_b$ for $b = a_{min}$ and $H_b = H^a_b$ for $a = a_{max}$. Set $\hat{\mathcal{H}}^a = \bigoplus_{\beta: a(\beta) \subseteq a} L^2(R^a_{a(\beta)})$.

b. Banach spaces

In the general case it is convenient to use spaces somewhat more sophisticated than $\mathcal{H}$-spaces introduced in section 4. We call, generically, these spaces the $\mathcal{B}$-spaces and define them as

$$\mathcal{B}_{\delta,\gamma}(R^a_b) = \bigcap_{a(\beta) \supseteq b} \chi_\beta(\gamma)^{-1}\left[\sum_{d \not\subseteq a(\beta)} J^\delta_{d(b)} L^2(R^a_b)\right] \quad \text{and} \quad \hat{\mathcal{B}}^a_{\delta,\gamma} = \bigoplus_\beta \mathcal{B}_{\delta,\gamma}(R^a_{a(\beta)}) .$$

The more general space R^a_b used in this definition allows to apply the $\mathcal{B}$-spaces inductively in the intermediate steps. The following embeddings follow from the definitions

$$L^2 \subseteq \mathcal{H}_{\delta,\gamma} \subseteq \mathcal{B}_{\delta,\gamma} \subseteq \mathcal{B}_{\delta',\gamma'} \subseteq L^2_\delta \quad \text{with } \gamma' \leq \gamma \text{ and } \delta' \geq \delta$$

The new property which we gain by introducing $\mathcal{B}$-spaces, as compared to the $\mathcal{H}$-spaces, is that the operators P_a are bounded on $\mathcal{B}_{\delta,\gamma}$ (see the end of appendix II). Here $P_a = E^a_d \otimes 1\!\!1_a$, where E^a_d is the eigenprojection on the discrete-spectrum subspace of H^a. This result is used in the inductive study of $L(z)$ and $F(z)$ on the $\mathcal{B}$-spaces (see appendix II).

c. Estimates near continuous spectrum

Now we proceed to the most difficult part of these lectures - estimating the resolvent near the continuous spectrum. To relieve the main text, the purely technical proofs are carried out into appendices II and III.

The central result of this section which implies the main theorem of section 4 (theorem 4.4 with condition (SR)) is the following:

Theorem 8.1. Let conditions (SR), (QB), (IE) be satisfied. Then for each $a \in A$, the resolvent of H^a has the form

$$R^a(z)(1\!\!1 - E^a_d) = J^a \hat{R}^a(z) Q^a(z), \tag{8.1}$$

where $Q^a(z)$ is an analytic in $z \in \bigcap_{b \subset a} \rho(H_b)$ family of bounded operators from $\mathcal{H}^a$ to $\hat{\mathcal{H}}^a$ which can be extended to bounded operators from $\mathcal{B}_{\delta,\gamma}(R^a)$ to $\hat{\mathcal{B}}^a_{\delta,\gamma}$, $\delta > 1$, $\gamma \in \kappa(\operatorname{Re} z)$, strongly continuous as $\operatorname{Im} z \to \pm 0$ and $\operatorname{Re} z \notin \sigma_p(H^a) \setminus \sigma_d(H^a)$ (the latter set is empty for $a \neq a_{max}$.

Proof. Henceforth we supress the superindex a and abreviate $\mathcal{B}_{\delta,\gamma} = \mathcal{B}_{\delta,\gamma}(R^a)$. All the operators and spaces appearing in this section have the superindex a.

We study the boundary values of the resolvent $R(z)$ using the equation

$$R(z)A(z) = F(z), \tag{8.2}$$

derived in section 5. The following propositions contain basic estimates on the operators $A(z)$ and $F(z)$.

Proposition 8.2. Under conditions (SR), (QB), (IE), the family $A(z)$, defined originally on $\mathcal{H}$ can be extended to a family of bounded operators on $\mathcal{B}_{\delta,\gamma}$, $\delta > 1$, $\gamma \in \kappa(\operatorname{Re} z)$, strongly continuous as $\operatorname{Im} z \to \pm 0$.

Proposition 8.3. Under conditions (SR), (QB), (IE) the family $F(z)$ can be represented as $F(z) = J\hat{R}(z)\hat{F}(z)$, where $\hat{F}(z)$ is an analytic in $z \in \bigcap_{b \subset a} \rho(H_b)$ family of bounded operators from H to $\hat{H}$ which can be extended to bounded operators from $\mathcal{B}_{\delta,\gamma}$ to $\hat{\mathcal{B}}_{\delta,\gamma}$, $\delta > 1$, $\gamma \in \kappa(\mathrm{Re}\, z)$, strongly continuous as $\mathrm{Im}\, z \to \pm 0$.

The proofs of these two propositions are given in appendix II. They are mainly combinatorial.

We define

$$L(z) = A(z) - \mathbb{1} .$$

Proposition 8.4. The family $[L(z)]^2$ is continuous in the uniform operator topology as $\mathrm{Im} z \to \pm 0$. Hence $[L(\lambda \pm i0)]^2$ are compact operators.

Before proceeding to the proof of proposition 8.4 we derive theorem 8.1 from propositions 8.2 - 8.4.. Since the potentials V_ℓ are dilation-analytic, the operator $L(z)$ and the dilation group $U(\rho)$ satisfy the restrictions of theorem 5.3. Therefore

$$\lambda \in \sigma_p(H) \Longleftrightarrow 0 \in \sigma(A(\lambda \pm i0)) \quad \text{for} \quad \lambda \notin \tau(H). \tag{8.3}$$

Note that this is the only place where we use the dilation-analyticity of the potentials.

Proposition 8.5. The following conditions are equivalent

(i) λ is a quasibound-state or usual eigenvalue of H.
(ii) $0 \in \sigma(A(\lambda \pm i0))$.

Proof. Let $A(z,\zeta)$ be the A-family for the delation-family $H(\zeta)$ (see section 5 for the definition of the A-families). $A(\lambda,\zeta)$ is defined and compact on $\mathcal{B}_{\delta,\gamma}$ with $\delta > 1$ and any γ. This can be shown by applying the machinery developed in this section but the latter is not necessary. The problem is essentially single-channel simplified by the fact that λ is a "semi-isolated" spectral point (λ lies on the tip of a branch of the continuous spectrum which is a semiline; see the figure in supplement II) and is tackled as in section 6. Moreover, one can easily show that $A(\lambda,\zeta)$ is analytic in $\zeta \in A$. Next, by the Combes argument (see the proof of lemma 5.4), the spectrum of $A(\lambda,\zeta)$ is locally ζ-independent. Hence

$$0 \in \sigma(A(\lambda \pm i0)) \Longleftrightarrow 0 \in \sigma(A(\lambda,\zeta)), \quad \zeta \in A \cap \mathbb{C}^{\pm} .$$

Furthermore, proposition 8.4, corollaries 6.8 and 5.9 and theorem 6.2 imply $0 \in \sigma(A(\lambda,\zeta)) \Longleftrightarrow$ H has a quasibound or bound state at λ.

The last two relations yield the statement of proposition 8.5. $\square$

Now we return to the derivation of theorem 8.1. We define $Q(z) = \hat{F}(z)A(z)^{-1}(\mathbb{1} - E_d)$. Propositions 8.2 - 8.5 **coupled with** conditions (SR), (QB), (IE) show that $Q(z)$ defines a family of bounded operators from $\mathcal{B}_{\delta,\gamma}$ to $\hat{\mathcal{B}}_{\delta,\gamma}$ with $\delta > 1$ and $\gamma \in \kappa(\mathrm{Re}\, z)$, which is strongly continuous as $\mathrm{Im}\, z \to \pm 0$ as long as $\mathrm{Re}\, z \notin \sigma_p(H)$. We have to show that the latter condition can be weakend to $z \notin \sigma_p(H) \setminus \sigma_d(H)$ (empty sets for $a \neq a_{max}$). First, recall that the operators $A(z)$ and $F(z)$ are built out of $R_b(z)$, $b \subsetneq a$, and the potentials. Since $\sigma_d(H) \subset \rho(H_b)$ $\forall\ b \subsetneq a$ (by condition (IE) and the HVZ theorem), they can be easily shown by methods of section 5 to depend analytically on z near $\sigma_d(H)$. Since $\mathrm{Ker}[A(z)^*] = \mathrm{Ker}(H-\bar{z})$ for $z \in \mathbb{C} \setminus \sigma_{ess}(H)$, $A(z)^{-1}(\mathbb{1}-E_d)$ is continuous at $\sigma_d(H)$. This implies the strong continuity of $Q(z)$ also at $\sigma_d(H)$, which completes the proof of theorem 8.1. $\square$

d. Proof of proposition 8.4. In appendix II we prove propositions 8.2 and 8.3 by the induction on the decompositions $a \in A$ (the superindex we omit!). We assume that they are true for all b with $b \subset a$. Then, by the derivation above, theorem 8.1 holds for all b with $b \subset a$ (b is the omitted superindex), i.e. $R^b(z)(\mathbb{1}^b - E^b_d)$ have the form $J^{b\hat{}b}R^b(z)Q^b(z)$ with $Q^b(z)$ described in the theorem. Furthermore, we have shown that $L(z)$ (and also $F(z)$) is a finite, linear combination of the terms which are products of the potentials V_ℓ, $\ell \subset a$, and the resolvents $R_b(z)$, $b \subset a$. Recall, that $R_b(z)$ and $R^b(z)$ are connected as $R_b(z) = \int R^b(z-s) \otimes \delta(T_b - s)ds$, where the integral can be understood by mapping it onto the direct fiber integral [RS1V] $\int^{\oplus} H_\lambda d\lambda$ with respect to T_b. Therefore, we can consider $L(z)$ (and also $F(z)$) as a function of the potentials V_ℓ, $\ell \subset a$, the eigenvectors, ψ^α, $a(\alpha) \subset a$, (coming from $R^b(z)E^b_d$) and families $Q^b(z)$, $b \subset a$, (coming from $R^b(z)(\mathbb{1} - E^b_d)$). Along the proof of propositions 8.2 and 8.3 we prove in appendix II the following

Lemma 8.6. Assume conditions (SR), (QB), (IE) are obeyed. Then the operators $L(z)$, considered as functions of the potentials $V_\ell \in L^p \cap L^q(R^\nu)$, $p > \nu/2 > q$, $\ell \subset a$, eigenvectors $\psi^\alpha \in L^2_\delta(R^{a(\alpha)})$, $\delta > 1$, $a(\alpha) \subset a$, and the operators $Q^b(z) \in \mathcal{L}_s(\mathcal{B}_{\delta,\gamma}(R^b), \hat{\mathcal{B}}^b_{\delta,\gamma})$, $\delta > 1$, $\gamma \in \kappa(\mathrm{Re}\, z)$, $b \subset a$, are norm continuous uniformly in $z \in \bigcap_{b \subset a} \rho(H_b)$. $\square$

Proof. See appendix II. $\square$

To fix ideas we consider below only the <u>upper half-plane $\mathbb{C}^+$</u>.

<u>Lemma 8.7.</u> Consider the family, $M(z)$, of operators obtained from $L(z)$ by replacing V^ℓ, $\ell \subseteq a$, and ψ^α, $a(\alpha) \subset a$, by C_0^∞ - functions and $Q^b(z)$, $b \subset a$, by integral operators with C_0^∞- kernels, infinitely and boundedly differentiable in $z \in \overline{\mathbb{C}^+}$. Then $M(z)^2$ is norm continuous (on $\mathcal{B}_{\delta,\gamma}$) as $\mathrm{Im}\, z \downarrow 0$.

To prove this lemma we use the complex distortion technique introduced below. Let

$$\varphi(t) = 1 \quad \text{for} \quad 0 \leq t \leq 1 \quad \text{and} \quad = t^{-2} \quad \text{for} \quad t \geq 1.$$

We define (in the <u>momentum representation</u>) the one-parameter family of unitary operators

$$U(\delta) : f(p) \to Cf(e^{\delta\varphi(|p|)}p), \tag{8.4}$$

where C is the normalizing factor:

$$C = [\text{Jacobian of the transformation } p \to e^{\delta\varphi(|p|)}p]^{1/2} . \tag{8.5}$$

We have obviously

$$M_1(z)^2 = (U(-\delta)M_1(z)^*)^* U(\delta)M_1(z).. \tag{8.6}$$

<u>Theorem 8.8.</u> For any $\varepsilon > 0$, there is a uniformly bounded family $M_1(z)$ s.t. $\|M(z)-M_1(z)\| \leq \varepsilon$ and $U(\delta)M_1(z)$ and $U(-\delta)M_1(z)^*$ have analytic continuations in δ into a strip along $\mathbb{R}$ obeying $\mathrm{Im}\delta \cdot \mathrm{Im} z \leq 0$. These continuations define analytic in $z \in \bigcap_{b\subset a} \rho(H_b)$ families of compact operators, norm-continuous as $\mathrm{Im}\, z \to +0$ ($\mathrm{Im}\, \delta < 0$).

This theorem is proven in appendix III.

The tric performed by the complex-distortion family $U(i\delta)$ is that the operators of the type described in lemma 4.7 become, when complexly distorted, norm-continuous as $\mathrm{Im}\, z$ approaches 0 from an appropriate semiplane (in fact, their fibres are Hilbert-Schmidt and Hilbert-Schmidt continuous; see lemma A III.8). Since the <u>continuity type</u> of <u>these operators</u> <u>determines</u> the <u>continuity type of $L(z)$</u> (or, respectively, $M(z)$) (see appendix II), the norm-continuity of $U(\delta)M(z)$ and $U(-\delta)M(z)^*$ follows in the same way as the strong-continuity of $L(z)$ (see appendix II).

<u>Corollary 8.9.</u> The right-hand side of (8.6) has an analytic continuation in δ into a stripe along $\mathbb{R}$ with $\mathrm{Im}\,\delta\, \mathrm{Im}\, z \leq 0$ (so eqn (8.6) holds also for those δ's). Hence $M_1(z)^2$ and $M(z)^2$ are norm continuous as $\mathrm{Im} z \downarrow 0$ and compact.

Now we return to the proof of proposition 8.4. By lemma 8.6, $L(z)^2$ is a continuous function of the potentials $V^\ell \in L^p_\delta \cap L^q(R^\nu)$, $p < \nu < 2q$, $\delta > 1$ (entering explicitely see eqn (8.9)), bound states $\psi^\alpha \in L^2_\delta(R^{a(\alpha)})$, $\delta > 1$, and $Q^b(z) \in L^\infty(\overline{\mathbb{C}^+}, \mathcal{B}_s)$ with $\mathcal{B}_s = \mathcal{L}_s(\mathcal{B}_{\delta,\gamma}(R^b), \hat{\mathcal{B}}^b_{\delta,\gamma})$ and by lemma 8.7 and lemma 8.10 below, $L(z)^2$ is continuous as $\mathrm{Im}\, z \downarrow 0$ on the dense subset of these variables. Hence $[L(z)]^2$ is norm continuous as z approaches $\bigcup_{b\subset a} \sigma(H_b)$ from above. In the same way one considers the $\mathbb{C}^-$ - boundary values. □

Lemma 8.10. Let $z \in \mathbb{C}^+$. The family $Q^b(z)$ can be approximated in the strong operator topology and uniformly in z by families $Q^b_n(z)$ of operators with C_0^∞ integral kernels which are analytic and bounded in $z \in \overline{\mathbb{C}^+}$ together with all z- derivatives:

$$\sup_{z\in\mathbb{C}^+} \|(Q^b(z) - Q^b_n(z))f\|_{\hat{\mathcal{B}}^b_{\delta,\gamma}} \to 0 \quad \text{as} \quad n \to \infty \quad \text{for all} \quad f \in \mathcal{B}_{\delta,\gamma}(R^b).$$

The same is true for $z \in \mathbb{C}^-$.

Proof. Let T_n and $\hat{T}_n$ be two sequences of integral operators on $L^2(R^b)$ and $\oplus\, L^2(R^b_c)$, respectively, with C_0^∞ integral kernels and converging strongly to $\mathbb{1}^b$ and $\hat{\mathbb{1}}^b$ as $n \to \infty$ (e.g. the integral kernels of T_n can be constructed as $\chi_n(x)\chi_n(y)\delta_n(x-y)$ where $\chi_n, \delta_n \in C_0^\infty$ and $\chi_n \to$ identical 1 and δ_n is a δ- sequence). Let furthermore $\varepsilon_n \downarrow 0$. The operators

$$Q^b_n(z) = \hat{T}_n Q^b(z + i\varepsilon_n) T_n$$

obey all the requirements of the statement of the lemma. □

9. Non-Dilation-Analytic Potentials

In this section we describe the changes which ought to be made in the derivations of section 8 in order to prove our main result, theorem 8.5 (implying the asymptotic completeness) under conditions (SR'), (IE) and (QB), i.e. without assuming the dilation analyticity.

The dilation analyticity was used in section 8 only in the study of the homogeneous equation

$$A(\lambda \pm i0) f = 0 \tag{9.1}$$

(the upper index a is dropped everywhere in this section), namely, to demonstrate that

$$0 \in \sigma(A(\lambda \pm i0)) \Longleftrightarrow \lambda \in \sigma_p(H), \tag{9.2}$$

if λ is not a two-cluster threshold. We show now how to prove this statement without assuming dilation analyticity (i.e. with condition (SR') replacing (SR)). Since for a solution f of (9.1), $F(\lambda \pm i0)f$ is a generalized eigenfunction of H corresponding to λ, it suffices to show that $F(\lambda \pm i0)f \in \mathcal{D}(H)$ if λ is not a two cluster threshold. We concentrate our attention in this section on the case of λ outside of the threshold set of H.

We begin with a few general remarks.

Lemma 9.1. Let H be a self-adjoint operator in Hilbert space $\mathcal{H}$, let $F(z) : \mathbb{C} \setminus \mathbb{R} \to \mathcal{L}(\mathcal{H}, \mathcal{D}(H))$ and let there exist a Banach space $\mathcal{B} \subset \mathcal{H}$ such that $F(z)$ is bounded from $\mathcal{B}$ into $\mathcal{B}'$ uniformly in $z \in \mathbb{C} \setminus \mathbb{R}$ and such that $A(z) \equiv (H-z)F(z)$ is defined on $\mathcal{B}$ for all $z \in \mathbb{C} \setminus \mathbb{R}$ on a domain independent of z and strongly continuous as $\operatorname{Im} z \to 0$. Then, if f_0 is a solution of one of the equations

$$A(\lambda \pm i0) f_0 = 0, \tag{9.3}$$

it satisfies

$$\lim_{\varepsilon \to \pm 0} \sqrt{|\varepsilon|}\, \| F(\lambda + i\varepsilon) f_0 \| = 0.$$

Proof. Since $A(z)$ is strongly continuous as z approaches $\mathbb{R}$, we have

$$A(z) f_0 = (A(z) - A(\lambda + i0)) f_0 \to 0 \quad \text{as} \quad z \to \lambda + i0$$

in $\mathcal{B}$. Let

$$\psi_z = F(z) f_0 \in D(H).$$

It follows that

$$(H-z)\psi_z = A(z)f_0 \to 0 \quad \text{as} \quad z \to \lambda + i0$$

in $\mathcal{B}$. Taking the scalar product of this equation with ψ_z and the imaginary part of the result, one gets

$$-2\mathrm{Im}\, z\|\psi_z\|^2 = \mathrm{Im}\langle A(z)f_0,\psi_z\rangle \to 0, \qquad (z \to \lambda + i0).$$

The same holds for $\lambda - i0$. □

Corollary 9.2. Let H and $\hat{H}$ be self-adjoint operators on Hilbert spaces $\mathcal{H}$ and $\hat{\mathcal{H}}$, respectively, and let J be a surjection from $\hat{\mathcal{H}}$ to $\mathcal{H}$. Assume that the conditions of lemma 9.1 are satisfied for H and there is a Banach space $\hat{\mathcal{B}} \subset \hat{\mathcal{H}}$ such that (a) $\delta_\varepsilon(\hat{H}-\lambda)$ is weakly continuous in $L(\hat{\mathcal{B}},\hat{\mathcal{B}}')$ as $\varepsilon \downarrow 0$, (b) $\|J\hat{R}(\lambda+i\varepsilon)u\|^2 - \|\hat{R}(\lambda+i\varepsilon)u\|^2 \to 0$ as $|\varepsilon| \to 0$ and (c) $F(z)$ is representable as

$$F(z) = J\hat{R}(z)\hat{F}(z),$$

where $\hat{F}(z)$ are bounded operators from $\mathcal{B}$ to $\hat{\mathcal{B}}$, uniformly bounded in $z \in \mathbb{C}\setminus\mathbb{R}$. Then

$$\langle(\hat{H}-\lambda)\hat{F}(\lambda+i0)f, \quad \hat{F}(\lambda+i0)f\rangle = 0.$$

Now let $\int^{\oplus}\hat{\mathcal{H}}_\lambda d\sigma(\lambda)$ be a representation of $\hat{\mathcal{H}}$ as the direct integral with respect to the operator $\hat{H}$ and let $\Pi = \{\Pi_\lambda\}$ be a unitary operator $\hat{\mathcal{H}} \to \int^{\oplus}\hat{\mathcal{H}}_\lambda d\sigma(\lambda)$.

Corollary 9.3. If, in addition to the restrictions of corollary 9.2, the following condition is satisfied:

$$\hat{\mathcal{B}} \subset \{\hat{f} \in \hat{\mathcal{H}} : \left|\, \|\Pi_\lambda\hat{f}\|_{\hat{\mathcal{H}}_\lambda} - \|\Pi_\nu\hat{f}\|_{\hat{\mathcal{H}}_\nu}\, \right| \leq C|\lambda-\nu|^\alpha\}, \quad \alpha > \frac{1}{2}, \tag{9.4}$$

then for every solution f_0 of (9.3), $F(\lambda\pm i0)f_0 \in \mathcal{D}(H)$ and is therefore an eigenfunction of H corresponding to the eigenvalue λ.

It is not difficult to show that the Shrödinger operator H, the asymptotic Hamiltonian $\hat{H}$, the identification J and the $\mathcal{B}_{\delta,\gamma}$-spaces defined in section 8 obey conditions (a) and (b) of corollary 9.2. However $\mathcal{B}_{\delta,\gamma}$ does not satisfy condition (9.4) of corollary 9.3. So we construct new Banach spaces (a chain with the base in $\mathcal{B}_{\delta,\gamma}$.

$$\mathcal{B}_{s,\delta,\gamma}(R_d) = \{f \in \mathcal{B}_{\delta,\gamma}(R_d) : \int\|F^{-1}\Delta^s(h)Ff\|_{\mathcal{B}_{\delta,\gamma}(R_d)}|h|^{-3m}dh < \infty\}, \quad s > 0,$$

$$\mathcal{B}_{0,\delta,\gamma}(R_d) = \mathcal{B}_{\delta,\gamma}(R_d)$$

and

$$\hat{B}_{s,\delta,\gamma} = \underset{\alpha}{\oplus} B_{s,\delta,\gamma}(R_d),$$

where F is the Fourier transform, $m = \nu(\#(d) - \#(a))$ and $(\Delta^s(h)f)(p) = |h|^{-s}(f(p+h) - f(p))$.

A realization of $\hat{H}$ as the direct fiber integral $\int^{\oplus} \hat{H}_\lambda d\lambda$ with respect to $\hat{H}$ and a unitary operator $\Pi = \int^{\oplus} \Pi_\lambda d\lambda$ from $\hat{H}$ to $\int^{\oplus} \hat{H}_\lambda d\lambda$ were constructed in subsection 4e. Below we use without mentioning the notations from that subsection.

It follows from lemmas 3.18 and 3.21 that the operators $\Pi_\alpha(\lambda)$ are bounded from $\Sigma J^\delta_{b(d)} L^2(R_d)$ to $L^2(\Omega_d)$, $d = a(\alpha)$ and $\delta > 1$, uniformly in λ. However, the family $\Pi_\alpha(\lambda)$ is not Hölder continuous in λ.

Lemma 9.4. The family $\Pi_\alpha(\lambda)$ verifies the estimate

$$\|\Delta^{s'}_\lambda(h)\,\Pi_\alpha(\lambda) f\|_{L^2(\Omega_d)} \leq C\lambda^{-s/2}\|f\|_{B_{s,\delta,\gamma}(R_d)}, \qquad \gamma \in \kappa(\lambda),$$

for $\delta > 1$, $s' < s$ and $d = a(\alpha)$.

The proof of this lemma follows from corollary IV.I of appendix IV in [S2].

Thus the spaces $B_{s,\delta,\gamma}$ with $\gamma \in \kappa(\lambda)$, $\delta > 1$ and $s > 1/2$ satisfy condition (9.4) of corollary 9.3. The conditions of corollary 9.2 are, of course, obeyed as well.

It is not difficult to adapt the estimates of appendix II to the case of the new spaces. This complicates only the proofs of technical propositions AII.14 and AII.15. Details and generalizations can be found in [S2].

10. Instability of Quasibound-State and Embedded Eigenvalues

It is easy to show that in the two-body case the quasibound-state eigenvalues are unstable under small perturbations of the potentials. More precisely, recall that a two-body system described by $-\Delta+V$ with $V \in L^{\nu/2}(\mathbb{R}^\nu)$ has a (quasi) bound state at 0 iff $-\Delta\psi+V\psi = 0$ has a nontrivial solution in $(\Delta)^{-1}L^p(\mathbb{R}^\nu)$, $p < \nu/2$. The latter condition is clearly equivalent to $-1 \in \operatorname{spec}[V(-\Delta)^{-1}$ on $L^p(\mathbb{R}^\nu)$, $p < \nu/2]$, this implies that $-1 \notin \operatorname{spec}[(1+\varepsilon)V(-\Delta)^{-1}]$ for any sufficiently small $\varepsilon \neq 0$ which is equivalent to $-\Delta\psi + (1+\varepsilon)V\psi = 0$ having only the trivial solution on $(\Delta)^{-1}L^p(\mathbb{R}^\nu)$, $p < \nu/2$. One expects that this situation persists in the many-body case as well. Motivated by an analysis of simple models, such as the Friedrichs model [F,Howl-8,Bau,HS] and the tunnel effect, one also supposes that the eigenvalues embedded into the continuous spectra are also unstable under small changes of the potentials.

In this section we show that in the general case of the many-body systems, the quasibound-state eigenvalues and the eigenvalues embedded into the continuous spectrum are unstable under small changes of the potentials (the result about the nonthreshold eigenvalues requires some extra implicit condition). Thus one can state that these eigenvalues are absent for "almost all" potentials, i.e. conditions (IE) and (QB), formulated in the introduction, are "almost always" satisfied.

Henceforth H is an N-body Hamiltonian with real, dilation analytic potentials and W_ℓ are some real, dilation-analytic, pair potentials. We begin our discussion with the quasibound-states and threshold eigenvalues. Accordingly all the potentials are assumed to be short-range $(L^q \cap L^p_\delta(\mathbb{R}^\nu)$, $\delta > 1$, $p < \nu < 2q)$ (remember, that the notion of quasibound state makes sense only for the short-range potentials).

a. Instability of Quasibound-State and Threshold Eigenvalues

Theorem 10.1. Let λ be a threshold of $H : \lambda = \lambda^\beta$.

(i) If H has neither a bound nor quasibound state at λ, then this situation persists for $H(\varepsilon) = H + \varepsilon\Sigma W_\ell$ at the corresponding thresholds $\lambda_j(\varepsilon)$ $(\lambda_j(0) = \lambda)$, for $|\varepsilon| \leq \varepsilon_0$ with some $\varepsilon_0 > 0$.

(ii) If H has either a bound or quasibound state (or both) at λ, then there exists $\varepsilon_0 > 0$ and potentials W_ℓ so that $H_1(\varepsilon) = H + \varepsilon \sum_{\ell \not\subset b} W$ has neither a bound nor quasibound state at λ (which remains to be a threshold of $H_1(\varepsilon)$). Here $b = a(\beta)$.

Proof. We recall (theorem 6.2) that H has a bound or quasibound state at λ if and only if $0 \in \sigma(A(\lambda,\zeta))$ on $F(L^p \cap L^2)(R)$, $p > \nu/\nu-2$ (i.e. λ is a singular point of $A(\lambda,\zeta)$ corresponding to the eigenvalue 0). Remember, that $A(\lambda,\zeta)$ is the A-family, defined in section 5, for $H(\zeta)$, the dilation-family associated with H. Analogously, we define $A(\lambda,\zeta,\varepsilon)$ and $A_1(\lambda,\zeta,\varepsilon)$ for the operators $H(\zeta,\varepsilon) = H(\zeta) + \varepsilon\Sigma W_\ell(\zeta)$ and $H_1(\zeta,\varepsilon) = H(\zeta) + \varepsilon \sum_{\ell \not\supset b} W_\ell(\zeta)$, respectively. Again, $H(\varepsilon) = H + \varepsilon\Sigma W_\ell$ has a bound or quasibound state at ν iff $0 \in \sigma(A(\nu,\zeta,\varepsilon))$ and similarly for $H_1(\varepsilon)$. So we come to studying the families $A(\lambda,\zeta,\varepsilon)$ and $A_1(\lambda,\zeta,\varepsilon)$.

Let $\lambda_j(\varepsilon)$ be the threshold of $H(\varepsilon)$ coming from λ:

$$\lambda_j(0) = \lambda \quad \text{and} \quad \lambda_j(\varepsilon) \in \sigma_d(H^b(\varepsilon)) \quad \text{for} \quad \varepsilon > 0 \tag{10.1}$$

then by the analytic perturbation theory (see [K2]), $\lambda_j(\varepsilon)$ are analytic at $\varepsilon = 0$.

Lemma 10.2. $A(\lambda_j(\varepsilon),\zeta,\varepsilon)$ are norm continuous in ε at 0. $A_1(\lambda,\zeta,\varepsilon)$ is analytic in ε at $\varepsilon = 0$.

The proof of this lemma retraces the steps of the proof of theorem 6.1 with additional reciting the words "norm-continuous/analytic in ε" in the appropriate moments. It also helps to notice that $\lambda_j(\varepsilon) \in \rho(H^c(\zeta,\varepsilon))$ for $c \not\supset b$ provided λ is a nondegenerate threshold (i.e. $\lambda^\gamma \neq \lambda$ for $a(\gamma) \not\supseteq b$) (and hence so are $\lambda_j(\varepsilon)$ for sufficiently small ε) which is clearly a generic situation. Hence, in this case, $R_c(\lambda_j(\varepsilon),\zeta,\varepsilon)$ is analytic in ε at 0. Also $R_b(\lambda_j(\varepsilon),\zeta,\varepsilon)\bar{P}_b(\zeta,\varepsilon)$ is analytic in ε at 0. Here $\bar{P}_b(\zeta,\varepsilon) = \mathbb{1} - P_b(\zeta,\varepsilon)$ with $P_b(\zeta,\varepsilon) =$ (total eigenprojection of $H^b(\zeta,\varepsilon)$ associated with the eigenvalues $\lambda_j(\varepsilon)) \otimes \mathbb{1}_b$. Note that b above is the same as in (10.1). Finally, it is useful to notice that condition (QB) implies that $0 \notin \sigma(A^c(\lambda,\zeta))$. The latter yields $0 \notin \sigma(A^c(\lambda_j(\varepsilon),\zeta,\varepsilon))$ for sufficiently small ε, which is needed for the induction (on $c \supseteq b$). Similarly for $A_1(\lambda,\zeta,\varepsilon)$.

Statement (i) of theorem 10.1 follows readily from lemma 10.2: the continuity of $A(\lambda_j(\varepsilon),\zeta,\varepsilon)$ at $\varepsilon = 0$ yields that

$$0 \notin \sigma(A(\lambda,\zeta)) \Rightarrow 0 \notin \sigma(A(\lambda_j(\varepsilon),\zeta,\varepsilon)) \quad \text{for } \varepsilon \text{ small enough,}$$

which we mentioned at the beginning of the proof, is equivalent to (i).

Statement (ii) follows from the following proposition by picking W_ℓ so that the numerator in (10.2) is not zero. □

Proposition 10.3. Any eigenvalue ν_ε of $A_1(\lambda,\zeta,\varepsilon)$, such that $\nu_0 = 0$ is differentiable at $\varepsilon = 0$ and

$$\nu'_\varepsilon\Big|_{\varepsilon=0} = \frac{<\varphi^*, W(\zeta)F(\lambda,\zeta)\varphi>}{<\varphi^*,\varphi>},$$

where $W(\zeta) = \sum_{\ell\in b} W_\ell(\zeta)$, φ(resp. φ^*) belongs to the null space of $A(\lambda,\zeta)$ (resp. $A(\lambda,\zeta)^*$). Here the numerator and, obviously, denominator and finite.

Proof. We consider here only real ν_ε which suffices for our purposes. In this case, ν_ε and the corresponding eigenfunction, φ_ε, are analytic at $\varepsilon = 0$ by the analytic perturbation theory (see [K2, p. 70]).

Before deriving (10.2) note that the finiteness of the numerator here is guaranteed by the restrictions on $W_\ell(\zeta)$ and lemmas 6.4 and 6.8. Furthermore, we observe that since $A(\lambda,\zeta)^* = F(\lambda,\zeta)^*(H(\zeta)^*-\lambda)$ and $\operatorname{Ker} F(\lambda,\zeta)^* = \{0\}$, the following holds

$$(H(\zeta)^* - \lambda)\varphi^* = 0 \quad \text{for} \quad \varphi^* \in \operatorname{Ker} A(\lambda,\zeta)^*. \tag{10.3}$$

Now to derive (10.2) we differentiate

$$(H_1(\zeta,\varepsilon) - \lambda)F(\lambda,\zeta,\varepsilon)\varphi_\varepsilon = \nu_\varepsilon\varphi_\varepsilon$$

at $\varepsilon = 0$. w.r. to ε, multiply the result scalarly by φ^* and use (10.3). □

Corollary 10.4. There exists an open set, V, of the coupling constants such that 0 belongs to its boundary and such that $H(\varepsilon) = H + \Sigma\varepsilon_\ell W_\ell$, where $\varepsilon = (\varepsilon_\ell)$, has no quasibound and bound states at its thresholds for all ε from V.

Corollary 10.5. There exists an open set, V, of the coupling constants such that 0 belongs to its boundary and such that $H(\varepsilon) = H + \Sigma\varepsilon_\ell W_\ell$, has property : no subsystem (i.e. $(H^C(\varepsilon))$ has a quasibound and/or bound state at its thresholds, for all $\varepsilon \in V$.

b. Instability of Non-Threshold Embedded Eigenvalues

Theorem 10.6. Let λ_0 be a non-threshold eigenvalue of H, embedded into its continuous spectrum. Let P be the eigenprojection associated with λ_0, $\bar{P} = 1\!\!1 - P$ and ψ^j, the eigenfunctions of H corresponding to λ_0. If W satisfies

$$\operatorname{Im} < \bar{P}W\psi^j, (\bar{P}H\bar{P}-\lambda_0 - i0)^{-1}\bar{P}W\psi^j > \neq 0 \tag{10.5}$$

at least for one ψ^j, then $H(g) = H + gW$ has no eigenvalues in a vicinity of λ_0 for $g \neq 0$, sufficiently small.

Remarks 10.7. (a) Condition (8.5) is equivalent to the relation

$$\bar{\Pi}_{\lambda_0} \bar{P}W\psi^j \neq 0 \quad \text{at least for one } j. \tag{10.6}$$

Here $\{\bar{\Pi}_\lambda\}$ is a unitary map obeying $\bar{P}H\bar{P}\bar{\Pi}_\lambda = \lambda\bar{\Pi}_\lambda$ (see section 3).

(b) The eigenvalue λ_0 of H moves to the second Riemann sheet of $< u,(H(g)-z)^{-1}v >$ (for dilation analytic u and v) and becomes a resonance.

(c) We do not know how to verify condition (10.5) except for the simplest cases. However, the common sense suggests (i.e. look at the Fourier transform) that it is satisfied for "almost all" potentials.

The proof of theorem 10.6 is based on the following abstract result.

Theorem 10.8. Let $H(g)$ be an analytic at $g = 0$ family of self-adjoint operators having a common domain and the following properties

(i) $H(0)$ has an eigenvalue λ_0 embedded into its continuous spectrum.
Denote by P the eigenprojection for $H(0)$ associated with λ_0 and $\bar{P} = \mathbb{1} - P$.

(ii) $\bar{P}H(g)\bar{P}$ has no eigenvalues in a neighbourhood of λ_0 for sufficiently small g.

(iii) The family

$$B(\lambda,g) = \Gamma^*(g)(\bar{P}H(g)\bar{P}-\lambda-i0)^{-1}\Gamma(g),$$

where $\Gamma(g) = g^{-1}\bar{P}H(g)P \equiv g^{-1}\bar{P}(H(g)-H(0))P$, is defined and analytic in (λ,g) at $(\lambda_0,0)$.

(iv) $\text{Im}\, B(\lambda_0,0) \neq 0$. (Here, recall $\text{Im}\, A = \frac{1}{2i}(A-A^*)$ on $\mathcal{D}(A)\cap\mathcal{D}(A^*)$).

Then for g sufficiently small but nonzero, the total multiplicity of the eigenvalues of $H(g)$ coming from λ_0 is strictly less than the multiplicity of λ_0.

Derivation of theorem 10.6 from theorem 10.8. Show that all the conditions of theorem 10.8 are satisfied:

(i) This is guaranteed by the conditions of theorem 10.6.

(ii) By the definition of $\bar{P}$, $\bar{P}H(0)\bar{P}$ has no eigenvalues in a neighbourhood of λ_0. Hence λ_0 is a resolvent point of the dilation family $\bar{P}(\zeta)H(\zeta,0)\bar{P}(\zeta)$, by a trivial extension of the Balslev-Combes theorem (see supplement II). Hence the Kato continuity-of-spectrum theorem [K2, p. 208] yields that $\lambda_0 \in \rho(\bar{P}(\zeta)H(\zeta,g)\bar{P}(\zeta))$ for g sufficiently small which proves (ii).

(iii) We write using the Combes extension (see supplement II, eqn (*)) with $\text{Im}\,\theta \gtrless$

$$B(\lambda,g) = \Sigma|\psi^j >< W\psi^j, \bar{P}(\bar{P}H(g)\bar{P}-\lambda \mp i0)^{-1}\bar{P}W\psi^j >< \psi^j|$$

$$= \Sigma < \bar{P}(\bar{\theta})W(\bar{\theta})\psi^j(\bar{\theta}),(\bar{P}(\theta)H(\theta,g)\bar{P}(\theta)-\lambda)^{-1}\bar{P}(\theta)W(\theta)\psi^j(\theta) >|\psi^j>< \psi^j| \quad ,$$

which shows (iii) (remember that, as was noticed in the previous paragraph, $\lambda \in \rho(\bar{P}(\theta)H(\theta,g)\bar{P}(\theta))$ for g_0 sufficiently small and λ close to λ_0).

(iv) This is translated readily into (10.5).

Thus all the conditions of theorem 10.8 are obeyed. By the continuity argument, the conditions are also satisfied for $H(g)$ near g_0 which is sufficiently close to 0 and any eigenvalue of $H(g)$ near λ_0. Then theorem 10.8 implies that the total multiplicity of the embedded eigenvalues of $H(g)$ near λ_0 for g small enough if no-zero varies continuously with g which is impossible. Theorem 10.6 is proven. □

c. Feshbach Projection Method, Proof of Theorem 10.8.

For a self-adjoint operator H and an orthogonal projection P we define the following families (here $\bar{P} = \mathbb{1} - P$)

$$B(\lambda) = PH\bar{P}(\bar{P}H\bar{P}-\lambda-i0)^{-1}\bar{P}HP, \tag{10.7}$$

$$A(\lambda) = PHP - \text{Re}\, B(\lambda), \quad \text{where } \text{Re}\, A = \frac{1}{2}(A + A^*) \text{ on } \mathcal{D}(A) \cap \mathcal{D}(A^*), \tag{10.8}$$

finite-rank families, and

$$\eta(\lambda) = \mathbb{1} - \bar{P}(\bar{P}H\bar{P}-\lambda-i0)^{-1}\bar{P}HP. \tag{10.9}$$

Lemma 10.9. If $B(\lambda_0)$ is defined on $P\mathcal{H}$ for some point λ_0, then

$$\text{Ker}(H-\lambda_0) = \eta(\lambda_0)[\text{Ker}(A(\lambda_0)-\lambda_0) \cap \text{Ker}(\text{Im}\, B(\lambda_0))] \cap \mathcal{D}(H)$$

(note that $\text{Ker}\ \eta(\lambda) = \{0\}$.).

Remark 10.10. As was mentioned above, the equation $[\text{Im}\, B(\lambda_0)]f = 0$ is equivalent to the statement that the vector $\bar{P}HPf$ evaluated in the $\bar{P}H\bar{P}$-representation vanishes at λ_0. So the requirement that $\eta(\lambda_0)f \in \mathcal{D}(H)$ for f satisfying $[\text{Im}\, B(\lambda_0)]f = 0$ can be replaced by a smoothness condition on $\bar{P}HPf$ in the $\bar{P}H\bar{P}$-representation.

Proof. Applying the projections P and $\bar{P}$ to the equation $(H-\lambda_0)\psi = 0$ and excluding $\bar{P}\psi$ we find

$$(PHP - B(\lambda) - \lambda_0)f = 0 \tag{10.10}$$

with $f = P\psi$. Since $\text{Im}\, B(\lambda) = PH\bar{P}\delta(\bar{P}H\bar{P}-\lambda)\bar{P}HP \geq 0$ for real λ's, the latter equation in equivalent to

$$(A(\lambda_0) - \lambda_0)f = 0 \tag{10.11}$$

and

$$[\text{Im}\, B(\lambda_0)]f = 0. \tag{10.12}$$

Furthermore, $\eta(\lambda_0)f = \psi \in \mathcal{D}(H)$.

Conversely, if f satisfies (10.11) and (10.12), then it obeys also (10.10). So $\psi = \eta(\lambda_0) f$ is a weak solution of $(H-\lambda_0)\psi = 0$. □

Now we derive theorem 10.8 from this lemma, applied to the family $H(g)$. Let $A(\lambda,g)$ be the A-family for $H(g)$, defined in (10.8). By lemma 10.9 all the eigenvalues, $\lambda^j(g)$, of $H(g)$ near λ_0 satisfy the equation

$$\det(A(\lambda,g) - \lambda) = 0 \tag{10.13}$$

and they are analytic functions in broken powers of g. Since for real g, $\lambda^j(g)$ are eigenvalues of a self-adjoint operator, $H(g)$, they are, in fact, analytic in g. Furthermore, they obey

$$\lambda^j(0) = \lambda_0. \tag{10.14}$$

Again by lemma 10.9 and the analytic perturbation theory (see [K2]), for each j there are m_j eigenfunctions $f_m(g)$ obeying

$$A(\lambda^j(g),g)f_m^j(g) = \lambda^j(g)f_m^j(g)$$

and

$$[\mathrm{Im}\, B(\lambda^j(g),g)]f_m^j(g) = 0. \tag{10.15}$$

Since $\lambda^j(g)$ are analytic, so are these eigenfunctions. Besides, $f_m^j(g)$ are linearly independent since the matrix $[\langle f_m^j(g), f_n^k(g)\rangle]$ is close to the identity matrix for small g.

On the other hand, since $B(\lambda,g)$ is continuous, condition (iv) implies that $\mathrm{Im}\, B(\lambda,g) \neq 0$ for (λ,g) sufficiently close to $(\lambda_0,0)$. Thus $\mathrm{Im}\, B(\lambda^j(g),g) \neq 0$ for g small enough. This together with (10.15) implies that the number of eigenfunctions $f_m^j(g)$, i.e. Σm_j, is strictly less than dim $P\mathcal{H}$, the dimension of the space which $\mathrm{Im}\, B(g,\lambda)$ acts on. This yields by lemma 10.9 that the total multiplicity of the eigenvalues of $H(g)$ near λ_0 for g sufficiently small but nonzero is strictly less than the multicity of the eigenvalue λ_0 of $H(0)$. □

Appendix I: Representations for L(z) and F(z)

In this appendix we study in detail the structure of the operators A(z) and F(z). The formulae obtained here are used in appendix II where the boundedness and norm continuity of these operators are proven. Notice a not quite usual labeling of the formulae below: They are labeled sometimes also by values of parameters they contain. So, for instance, (AI.6b) stands for $L_b = \sum L_{f_1} B_{f_2} \cdots B_{f_t}$.

First we recall the main definitions (see section 5):

$$A_a(z) = (H_a - z)(H_0 - z)^{-1} \overrightarrow{\prod_{b \subset a}} A_b(z)^{-1}, \tag{AI.1}$$

and

$$F_a(z) = (H_0 - z)^{-1} \overrightarrow{\prod_{b \subset a}} A_b(z)^{-1}. \tag{AI.2}$$

These families obey

$$A_a(z) = (H_a - z) F_a(z). \tag{AI.3}$$

Now we introduce the operators

$$L_a(z) = A_a(z) - \mathbb{1} \quad \text{and} \quad B_a(z) = A_a(z)^{-1} - \mathbb{1} \tag{AI.4}$$

Using these definitions we compute

$$B_a(z) = -L_a(z) A_a(z)^{-1} = -L_a(z) F_a(z)^{-1} R_a(z). \tag{AI.5}$$

Lemma AI.1. The operators $L_a(z)$ can be expressed as follows (the argument z is suppressed)

$$L_a = \Sigma\, L_{f_1} B_{f_2} \ldots B_{f_t}, \tag{AI.6a}$$

where the sum runs over (not all!) partitions $(f_1, f_2, \ldots, f_t)$ satisfying

$$f_1 \neq f_2 \neq \ldots \neq f_t, \quad \#(f_1) = \#(f_2) \geq \ldots \geq \#(f_t) > \#(a)+1 \tag{AI.7a}$$

and

$$\bigcup_1^{s-1} f_i \not\subseteq f_s, \quad s = 2, \ldots, t, \quad \bigcup_1^{t} f_i = a \tag{AI.8a}$$

Proof. First we mention one operator identity which plays the key role in derivation of (AI.6a). Let L_i, $i = 1, \ldots, s$, be bounded operators such that $\mathbb{1} + L_i$, $i = 1, \ldots, s$, and $\mathbb{1} + \sum_{i=1}^{s} L_i$ have bounded inverses. Then

$$(\mathbb{1} + \sum_{i=1}^{s} L_i) \overrightarrow{\prod_{i=1}^{s}} (\mathbb{1} + L_i)^{-1} = \sum_{\substack{s \geq r \geq 1 \\ i_0 > i_1 > \ldots > i_r}} L_{i_0} B_{i_1} \ldots B_{i_r}, \tag{AI.9}$$

where $B_i = (\mathbb{1} + L_i)^{-1} - \mathbb{1}$. The derivation of this identity elementary: one multiplies $(\mathbb{1} + \Sigma L_j)$ by $(\mathbb{1} + L_i)^{-1}$ one by one. For instance on the first step we have

$$(\mathbb{1} + \sum_1^s L_i)(\mathbb{1} + L_1)^{-1} = \mathbb{1} + \sum_2^s L_i + \sum_2^s L_i B_1$$

and so on.

Now we show by induction in s that

$$(H_a - z) \prod_{i=s+1}^{\overrightarrow{N}} \prod_{\substack{f \subseteq a \\ \#(f)=i}} A_f^{-1} = \mathbb{1} + \sum_{\substack{c \subseteq a \\ \#(b)=s}} L_b + C_s , \tag{AI.10s}$$

where the orders in products $\prod_{\substack{f \subseteq a \\ \#(f)=i}} A_f^{-1}$ are arbitrary but fixed once and for all, $A_f(z) = H_0 - z$ for $\#(f) = N$, the operators L_b have representation (AI.6b) - (AI.8b) and C_s is an operator of the form

$$C_s = \sum L_f B_g \ldots B_c \tag{AI.11s}$$

with the partitions in the sum satisfying

$$f \neq g \neq \ldots \neq c, \quad \#(f) = \#(g) \geq \ldots \geq \#(c) \geq s+1 \tag{AI.12s}$$

and

$$f \cup g \cup \ldots \cup c \subseteq a, \quad \#(f \cup g \cup \ldots \cup c) \leq s - 1 \tag{AI.13s}$$

$$\text{for} \quad s \geq \#(a)+1 \quad \text{and} \quad C_s = 0 \quad \text{for} \quad s = \#(a) \tag{AI.14s}$$

Equation (AI.11s)-(AI.14s) is obvious for $s = N-1$, it follows from the definition $L_f = V_f R_0$, $\#(f) = N-1$. On the other hand equations (AI.10s), (AI.14s) with $s = \#(a)$ is exactly (AI.6a)-(AI.8a).

Assuming (AI.10s+1)-(AI.14s+1) and taking into account (AI.9) we find

$$(H_a - z) \prod_{i=s+1}^{\overleftarrow{N}} \prod_{\substack{f(a) \\ \#(f)=i}} A_f^{-1} = \mathbb{1} + \sum_{\substack{b_{i_k} \subset a \\ \#(b_{i_k})=s+1}} L_{b_{i_0}} B_{b_{i_1}} \ldots B_{b_{i_r}} + C_{s+1} \prod_{\substack{b \subseteq a \\ \#(b)=s+1}} (\mathbb{1} + B_b),$$

$$i_0 > i_1 < \ldots < i_r, \quad 1 \leq r \leq n_{s+1} \tag{AI.15}$$

where n_i is the number of elements in $A_i \equiv \{a \in A, \ \#(a) = i\}$. Taking into account (AI.11s+1) we regroup the r.h.s. of (AI.15) so it takes the form of (AI.10s)-(AI.14s). Namely, for each b with $\#(b) = s$, we collect b-connected terms

on the r.h.s. of (AI.15). e.g.
$L_{b_{i_0}} B_{b_{i_1}} \dots B_{b_{i_r}}$, $b_{i_0} \cup \dots \cup b_{i_r} = b$, into L_b-operators and throw all the other terms, i.e. those of higher connections, into C_s. □

<u>Lemma AI.2</u>. The operator $F_a(z)$ is the sum of disconnected and connected terms of the form,

$$F_a(z) = D_a(z) + C_a(z), \tag{AI.16}$$

where

$$D_a(z) = \sum_{b \subsetneq a} (R_b(z) - D_b(z)) \quad \text{and} \quad D_a(z) = 0 \text{ if } \#(a) = N \tag{AI.17}$$

and

$$C_a(z) = \Sigma\, R_{f_1}(z) L_{f_1}(z) B_{f_2}(z) \dots B_{f_t}(z) \tag{AI.18}$$

with $f_1, \dots, f_t$ satisfying (AI.7a, 8a).

<u>Proof</u>. Using definition (AI.2) of $F_a(z)$, substituting $A_c^{-1} = 1\!\!1 + B_c$ in there and expanding the product we arrive at

$$F_a = R_0 \sum_{f \cup \dots \cup g \subseteq a}'' B_f \dots B_g\,, \tag{AI.19a}$$

where the double prime means that the summation is performed only over those monomials which appear in the expansion of

$$\prod_{f \subsetneq a} (1\!\!1 + B_f).$$

We split sum (AI.19a) into two terms: the sum, $D_a(z)$, of all a-disconnected terms (i.e. $f \cup \dots \cup g \subsetneq a$) and the sum, $C_a(z)$, of all a-connected terms (i.e. $f \cup \dots \cup g = a$).

We rearrange the summation in the first sum ($D_a(z)$) in the following way

$$D_a = R_0 \sum_{b \subsetneq a} \sum_{f \cup \dots \cup g = b}'' B_f \dots B_g =$$

$$= R_0 \sum_{b \subsetneq c} \left\{ \sum_{f \cup \dots \cup g = b}'' B_f \dots B_g + \sum_{f \cup \dots \cup \varphi \subseteq b}'' B_f \dots B_\varphi\, B_b \right\}$$

$$= \sum_{b \subsetneq a} \left\{ C_b + F_b\, B_b \right\}. \tag{AI.20}$$

On the other hand, equation (AI.3) and the definition of B_a imply

$$R_a = F_a A_a^{-1}$$

$$= F_a(\mathbb{1} + B_a). \tag{AI.21}$$

This equation along with (AI.20) produces

$$D_a = \sum_{b \subsetneq a} (R_b - D_b) \ ,$$

which is the desired expression for $D_a(z)$.

Now we transform the operator $C_a(z)$. Equation (AI.19b) together with the equation $B_b = - F_b^{-1} R_b L_b$ implies

$$R_0 \sum_{f \cup \ldots \cup \varphi = \psi}'' B_f \ldots B_\varphi B_\psi = - R_\psi L_\psi \ .$$

Applying these equations (with different ψ) to the expression $C_a = R_0 \sum_{f \cup \ldots \cup g = a} B_f \ldots B_g$ we obtain (AI.18) with (AI.7a, 8a). □

Appendix II. Boundary Values of L(z) and F(z) (Proof of propositions 8.8 and 8.9)

We prove propositions 8.2 and 8.3 by induction on the partitions $a \in A$ (the superindex a is omitted in section 8). It is convenient to prove a statement slightly more general than those propositions, which is suited better for the induction proof. Instead of the operators $F^a(z)$ and $L^a(z)$ we consider the operators $F_b^a(z)$ and $L_b^a(z)$, respectively Here, of course, $b \subset a$. The latter operators are obtained from $F_b(z)$ and $L_b(z)$ by restricting them to the space $L^2(R^a)$. This restriction amounts, practically, to the substruction T_a from all the Hamiltonians involved, e.g. $H_b^a = T^a + \sum_{\ell \subseteq b} V_\ell$ on $L^2(R^a)$. It was used without mentioning by name in section 7. Note that $F_a^a = F^a$ and $L_a^a = L^a$. **We study the operators** F_b^a and L_b^a by the induction on $b \subset a$ for all possible a simultaneously. So we break the promise given at the very beginning of this appendix to conduct the induction on a. To keep up with the promise we would have to change the notation: $b \leftrightarrow a$. We use the notation above in order to keep up with the notations of section 8. We use the abbreviation $\mathcal{B}_\gamma^a \equiv \mathcal{B}_{\delta,\gamma}(R^a)$. The index $\delta > 1$ is fixed throughout the appendix. In what follows, we omit the superindex a. Again, later the reader should remember that we work on $L^2(R^a)$ with operators and spaces which, in general, have the superindex a.

We modify and adopt a piece of terminology from G. Hagedorn [Ha1].

Definition AII.1. Let A(z) be a family of operators from X to Y, two Banach spaces, defined for $z \in \mathbb{C} \setminus \mathbb{R}$ and let A(z) depend explicitly on the potential V^ℓ and the resolvents $R^b(z)$ (in this case it also depends implicitly on V^ℓ through $R^b(z)$). We say that A(z) is a well-behaved family of operators from X to Y iff it is strongly continuous as $\operatorname{Im} z \to \pm 0$ and together with its boundary values on $\mathbb{R}$ is norm continuous as a function of the potentials $V^\ell \in L_\delta^q \cap L^p(\mathbb{R}^\nu)$, $\delta > 1$, $p < \nu < 2q$, the eigenfunctions $\psi^\alpha \in L_\delta^2(R^{a(\alpha)})$, $\delta > 1$, and the operators $Q^b(z) \in L_s(\mathcal{B}_\gamma^b, \mathcal{B}_\gamma^b)$, $\gamma \in \kappa(\operatorname{Re} z)$, uniformly on compact subset of $\overline{\mathbb{C}^+}$ and $\overline{\mathbb{C}^-}$.

In order to make this appendix more autonomous of the main part of the text we remind here some definitions from section 4: c(d) stands for the pair (d,c) s.t. $d \subset c$ and $\#(d) = \#(c)+1$. With each pair c(d) we associate the multiplication operator

$$J_{c(d)}^\delta : u(x) \to (1+|x_d^c|^2)^{-\delta/2} u(x)$$

Furthermore

$\chi_\alpha(\gamma) = \chi(H_\alpha - \gamma)$ if $a(\alpha) \neq a_{min}$ and $\chi_\alpha(\gamma) = 1$ if $a(\alpha) = a_{min}$.

Here χ is a smeared out step function:

$$\chi \in C^{\infty}(\mathbb{R}), \quad \chi(t) = 1 \text{ if } t > -\frac{1}{10}\kappa \text{ and } \chi(t) = 0 \text{ if } t < -\frac{1}{5}\kappa$$

To realize the induction argument in a smooth way we introduce in addition to ($\delta > 1$ is fixed and omitted)

$$\mathcal{B}_{\gamma}(R_d) = \bigcap_{a(\beta)\supseteq d} \chi_{\beta}(\gamma)^{-1}[\sum_{c\nsubseteq a(\beta)} J^{\delta}_{c(d)} L^2(R_d)], \tag{AII.1}$$

the following intermediate spaces

$$\begin{aligned}\mathcal{B}^{(c)}_{\gamma}(R_d) &= [\mathcal{B}_{\gamma-T_c}(R^c_d) \otimes L^2(R_c)] \\ &\cap [L^2(R^c_d) \otimes \mathcal{B}_{\gamma}(R_c)] \\ &\cap [\bigcap_{a(\sigma)=c} \chi_{\sigma}(\gamma)^{-1}[L^2_{\delta}(R^c_d) \otimes L^2(R_c)] \quad \text{with } d\subset c\end{aligned} \tag{AII.2}$$

The first symbol on the r.h.s. of this equation is used somewhat loosely. Its meaning can be read from

$$\mathcal{B}_{\gamma-T_c}(R^c_d) \otimes L^2(R_c) = \bigcap_{\beta:d\subseteq a(\beta)\subset c} \chi_{\beta}(\gamma)^{-1} [\sum_{\substack{e\subseteq a(\beta)\\ e\subset c}} J^{\delta}_{e(d)} L^2(R_d)] .$$

We retain it since it reflects the essential structure of the space $\mathcal{B}^{(c)}_{\gamma}(R_d)$. Furthermore we define

$$\hat{\mathcal{B}}^{(c)}_{\gamma} = \oplus \ \mathcal{B}^{(c)}_{\gamma}(R_d),$$

where sum extends first over all channels σ with $a(\sigma) = d$ and then over all d satisfying $d\subset c$. We also use the shorthand

$$\mathcal{B}^{(c)}_{\gamma} = \mathcal{B}^{(c)}_{\gamma}(R).$$

and the notation $\mathcal{B}_{\lambda}$ for any of the space $\mathcal{B}_{\gamma}$, $\gamma \in \kappa(\lambda)$, where, recall, $\kappa(\lambda) = (\lambda-\frac{1}{2}\kappa, \lambda+\frac{1}{16}\kappa)$, $\kappa = \min\limits_{a(\beta)\subset a(\alpha)}(\lambda^{\beta}-\lambda^{\alpha})$. Here both symbols might be simultaneously decorated by other indices.

The new intermediate spaces are related to the $\mathcal{B}$-spaces as

$$\mathcal{B}^{(c)}_{\gamma}(R_d) \subset \mathcal{B}_{\gamma}(R_d) \text{ and } \hat{\mathcal{B}}^{(c)}_{\gamma} \subset \hat{\mathcal{B}}_{\gamma} \tag{AII.3}$$

These relations will be proved in lemma AII.20 at the end of this appendix.

Now we can proceed to the main result of this appendix. In what follows $\lambda = \mathrm{Re}\, z$. Set $J_b(\oplus u) = \sum_{a(\alpha) \subsetneq b} J_\alpha u_\alpha$ and $P_b = \sum_{a(\beta)=b} P_\beta$.

Proposition AII.3. Let conditions (SR), (QB), (IE) be satisfied. Then for any b and a with $b \subset a$ (remember, the superindex a is omitted!)

(i) The family $F_b(z)$ is representable as $F_b(z) = J_b\hat{R}(z)\hat{F}_b(z)$, where $\hat{F}_b(z)$ is an analytic in $z \in \bigcap_{c \subset b} \rho(H_c)$ family of bounded operator from $\mathcal{H}$ to $\hat{\mathcal{H}}$ which can be extended to a well-behaved family from $\mathcal{B}_\lambda$ to $\hat{\mathcal{B}}_\lambda$.

(ii) The families $F_b(z)^{-1} J_b \hat{R}(z)$ and $F_b(z)^{-1} P_\beta$, $a(\beta) = b$, can be extended to well-behaved families of operators from $\hat{\mathcal{B}}_\lambda$ to $\mathcal{B}_\lambda$ and from $\mathcal{B}_\lambda$ to $\mathcal{B}_\lambda$, respectively.

(iii) $L_b(z)$ can be extended to a well-behaved family of operators from $\mathcal{B}_\lambda$ to $\mathcal{B}_\lambda^{(b)}$.

Remarks AII.4. (a): (i) and the equation

$$L_b(z) = (H_b - z) F_b(z) \tag{AII.4}$$

imply that $L_b(z)$ is well-behaved on $\mathcal{B}_\lambda$. Indeed, using that for any $c \subset b$,

$$H_b = H^c \otimes \mathbb{1}_c + \mathbb{1}^c \otimes T_c + I_{c \setminus b} \quad \text{with} \quad I_{c \setminus b} = \sum_{\substack{\ell \subset b \\ \ell \not\subset c}} V_\ell , \tag{AII.5}$$

we can write

$$(H_b - z) F_b(z) = J_b \hat{F}_b(z) + \sum_{a(\sigma) \subset b} I_{a(\sigma) \setminus b} J_\sigma R_\sigma(z) F_{b,\sigma}(z),$$

where $\{F_{b,\sigma}\} = \hat{F}_b$. Using lemma AII.15 and the fact that $\psi^\sigma u \in \mathcal{B}_\lambda$ for $u \in \mathcal{B}_\lambda(R_{a(\sigma)})$ we obtain the desired result. Hence, all that (iii) actually tells us is that the range of $L_b(z)$ lies in a smaller space $\mathcal{B}_\lambda^{(b)} \subset \mathcal{B}_\lambda$.

(b) Propositions 8.2 and 8.3 follow from proposition AII.3.

Proof of proposition AII.3. The proof is conducted by the induction on the partitions $b \in \mathcal{A}$ for all different a, $a \supseteq b$, simultaneously. The statement is trivial for $b = a_{min}$.

We assume that the statement of proposition AII.3 holds for all pairs (c,a) with $c \subset a$ and $c \subset b$ and we prove it for all (b,a) with different a, $a \supseteq b$, (b is fixed). To this end we use equations (AI.2), (AI.6) and (AI.22) which show that the operators $F_b(z)$, $F_b(z)$ and $L_b(z)$ are built out of the operators $L_c(z)$ and $B_c(z)$ with different $c \subset b$, $c \neq b$. The information about $L_c(z)$ is obtained from the induction hypothesis. Our next task is to study $B_c(z)$

Using the definition $B_c(z) = (\mathbb{1} + L_c(z))^{-1} - \mathbb{1}$ and the basic equation $R_c(z)(\mathbb{1} + L_c(z)) = F_c(z)$ we derive

$$B_c(z) = - L_c(z)(\mathbb{1} + L_c(z))^{-1} = - L_c(z)F_c(z)^{-1}R_c(z). \tag{AII.6}$$

We write

$$R_c(z) = \sum R_\sigma(z)P_\sigma + R_c(z)(\mathbb{1} - P_c), \tag{AII.7}$$

where P_σ is the projection associated with the eigenfunction ψ^σ and $P_c = \sum_{a(\sigma)=c} P_\sigma$

Since for $a(\sigma) = c$, $R_\sigma(z)$ commutes with $L_c(z)$ and $F_c(z)$ we obtain that

$$B_c(z)P_c = \sum R_\sigma(z)K_\sigma(z), \tag{AII.8}$$

where

$$K_\sigma(z) = - L_c(z)F_c(z)^{-1}P_\sigma \tag{AII.9}$$

Lemma AII.5. The families $K_\sigma(z)$, defined by (AII.9), are well behaved on $\mathcal{B}_\lambda$.

Proof: The desired properties of $K_\sigma(z)$ follow from the induction hypothesis on $L_c(z)$ and $F_c(z)^{-1}P_\sigma$ (remember that $c \subset b$). □

Now we estimate $B_c(z)(\mathbb{1} - P_c)$. To this end we will first show that $R_c(z)(\mathbb{1} - P_c)$ admits a representation similar to one described in theorem 8.7. We begin with

Lemma AII.6. For each c with $c \subset b$, the family $[L^c(z)]^2$ is norm continuous as $\mathrm{Im}\, z \to \pm 0$. Hence its boundary values on $\mathbb{R}$ are compact.

Proof. The statement follows from the induction hypothesis on $L_c(z)$ with $a = c$ and lemma 8.7 (see the paragraph preceding lemma 8.10). □

Let E_d^c be the projection in $L^2(R^c)$ on the descrete spectrum eigenspace of H^c (so $P_c = E_d^c \otimes \mathbb{1}_c$). Recall, furthermore $\hat{R}^c(z) = (\hat{H}^c - z)^{-1}$ with $\hat{H}^c = \bigoplus_{a(\alpha)\subset c} (\lambda^\alpha + T^c_{a(\alpha)})$ and $J^c(\oplus u) = \sum_{a(\alpha)\subset c} J_\alpha u_\alpha$ on $\hat{\mathcal{H}}^c = \bigoplus_{\alpha: a(\alpha)\subset c} L^2(R^c_{a(\alpha)})$.

Lemma AII.7. For each $c \subset b$, $R^c(z)(\mathbb{1} - E_d^c) = J^c\hat{R}^c(z)Q^c(z)$ where $Q^c(z)$ extends to a family of uniformly bounded operators from $\mathcal{B}_\lambda(R^c)$ to $\hat{\mathcal{B}}^c_\lambda$, which is strongly continuous as $\mathrm{Im}\, z \to \pm 0$.

Proof. By the induction hypothesis with $a = c$, $F^c(z) = J^c\hat{R}^c(z)\hat{F}^c(z)$, where $\hat{F}^c(z)$ has the same properties as described in the lemma for $Q^c(z)$, and $L^c(z)$ can be extended to a well-behaved family of operators on $\mathcal{B}_\lambda(R^c)$. Furthermore, by lemma AII.7, the family $[L^c(z)]^2$ is norm continuous as $\text{Im } z \to \pm 0$ and therefore is compact together with its boundary values on $\mathbb{R}$. Finally, it was shown in section 8 (proposition 8.5 and the paragraph thereafter) that $(\mathbb{1} + L^c(z))^{-1}(\mathbb{1} - E_d^c)$ is well-behaved. Hence the equation

$$R^c(z) + R^c(z)L^c(z) = F^c(z)$$

implies the statement of the lemma. □

Now we are prepared for the estimation of $R_c(z)(\mathbb{1}-P_c)$ which we need for the study of $B_c(z)(\mathbb{1}-P_c)$.

<u>Lemma AII.8.</u> For each c with $c \subset b$, the equation

$$R_c(z)(\mathbb{1} - P_c) = J_c\hat{R}(z)Q_c(z) \tag{AII.10}$$

holds with a family, $Q_c(z)$, of operators from $\mathcal{B}_\lambda$ to $\hat{\mathcal{B}}_\lambda$ uniformly bounded in $z \in \bigcap_{d \subseteq c} \rho(H_d)$ and strongly continuous as $\text{Im } z \to \pm 0$.

<u>Proof.</u> Let $c \subset b$. We define

$$Q_c(z) = \int Q^c(z-s) \otimes \delta(T_c-s)ds, \tag{AII.11}$$

where the integral is understood in the weak sense on nice functions or in the T_c-representation (the Fourier transform in R_c). We derive the estimates on $Q_c(z)$ in two steps.

The first step is the following

<u>Lemma AII.9.</u> Let $\mu^c = \min_{a(\rho) \subsetneq c} \lambda^\rho$ and an operator $G(z)$, $z \in \mathbb{C} \setminus [\mu^c, \infty)$, be bounded from $\mathcal{B}_\lambda(R^c)$ into $\hat{\mathcal{B}}_\lambda^c$ and be strongly continuous in $\mathbb{C} \setminus [\mu^c, \infty)$ up to the boundary $[\mu^c, \infty)$. If, moreover, $\partial_\lambda^k G(z)$, where $k = 0, \ldots [\delta]+1$, and $\lambda = \text{Re } z$, are bounded from $L^2_{\delta'}(R^c)$ into $\oplus L^2_{\delta'}(R^c_d)$, $0 \leq \delta' \leq \delta$, for $\lambda < \mu^c$ then the operator

$$G'(z) = \int G(z-s) \otimes \delta(T_c-s)ds, \; z \in \mathbb{C} \setminus [\mu^c, \infty)$$

is bounded from $\mathcal{B}_\lambda^{(c)}$ into $\hat{\mathcal{B}}_\lambda^{(c)}$ and strongly continuous in $z \in \mathbb{C} \setminus [\mu^c, \infty)$ up to the boundary $[\mu^c, \infty)$.

Proof: Performing the Fourier transform, F_{R_c}, in the variables of R_c, we get

$$F_{R_c} G'(z)f = G(z-\tau(p_c))(F_{R_c} f)(p_c), \tag{AII.12}$$

where we write $f(p_c)$ for $f \in L^2(R)$, when f is considered as a vector-function from R_c into $L^2(R^c)$. Here $\tau(p) = |p|'^2$.

We use the following equality

$$||u||_{\mathcal{B}_\gamma(R_c)} = ||F_{R_c} u||_{F_{R_c}\mathcal{B}_\gamma(R_c)}. \tag{AII.13}$$

(AII.12) and (AII.13) imply (remember the definition of $\chi_\beta(\gamma)$)

$$||G'(z)f||_{\hat{\mathcal{B}}^{(c)}_\gamma} \leq || \, ||G(z-\tau(\cdot))F_{R_c} f(\cdot)||_{\hat{\mathcal{B}}^c_{\gamma-\tau(\cdot)}} ||_{L^2(R_c)} +$$

$$\sum_{\beta, a(\beta)\supseteq c} || \, || \theta(\lambda^\beta+\tau(\cdot)-\gamma)G(z-\tau(\cdot))F_{R_c} u(\cdot)||_{\oplus L^2(R^c_d)} ||_{L^2(R_c)}$$

$$+ \sum_{a(\sigma)=c} || \, ||\chi(\lambda -\gamma+\tau(\cdot))G(z-\tau(\cdot))F_{R_c} f(\cdot)||_{\oplus L^2(R^c_d)} ||_{L^2(R_c)} \tag{AII.14}$$

where $u = (\sum_{g\nsubseteq a(\beta)} J^\delta_{g(c)})^{-1}\chi_\beta(\gamma)f$ and θ is the characteristic function of $[\frac{1}{5}\kappa,\infty)$.

Taking into account this estimate and using the condition $\lambda-\frac{1}{2}\kappa < \gamma$ we obtain

$$||G'(z)f||_{\hat{\mathcal{B}}^{(c)}_\gamma} \leq$$

$$\leq \sup_{s\geq 0} ||G(z-s)||_{\mathcal{B}_{\gamma-s}(R^c) \to \hat{\mathcal{B}}^c_{\gamma-s}} || \, ||f(\cdot)||_{\mathcal{B}_{\gamma-\tau(\cdot)}(R^c)} ||_{L^2(R_c)} +$$

$$+ \sup_{t<\mu^c-\frac{3}{10}\kappa} \sum_{k=0}^{[\delta]+1} ||\partial_t^k G(t+i\varepsilon)||_{L^2(R^c) \to \oplus L^2(R^c_d)} || \, ||u(\cdot)||_{L^2(R^c)} ||_{L^2(R_c)}$$

$$+ \sup_{t<\mu^c-\frac{3}{10}\kappa} ||G(t+i\varepsilon)||_{L^2_\delta(R^c) \to \oplus L^2_\delta(R^c_d)} \sum_{\sigma, a(\sigma)=c} || \, ||(\chi_\sigma(\gamma)f)(\cdot)||_{L^2(R^c)} ||_{L^2(R_c)}$$

where $\varepsilon = \operatorname{Im} z$.

Using (AII.2) we finally find

$$||G'(z)||_{\mathcal{B}^{(c)}_\gamma \to \hat{\mathcal{B}}^{(c)}_\gamma} \leq \sup_{s\geq 0} ||G(z-s)||_{\mathcal{B}_{\gamma-s}(R^c) \to \hat{\mathcal{B}}^c_{\gamma-s}} +$$

$$+ \sup_{\substack{t<\mu^c - \frac{3}{10}\kappa \\ 0\leq\delta'\leq\delta}} \sum_{k=0}^{[\delta]+1} ||\partial_t^k G(t+i\varepsilon)||_{L^2_\delta(R^c) \to \oplus L^2_\delta(R^c_d)} \qquad \text{(AII.15)}$$

Applying a similar procedure to the operator $G'(w)-G'(z)$ we conclude that $G(z)$ is strongly continuous, considered from $\mathcal{B}^c_\lambda$ to $\hat{\mathcal{B}}^{(c)}_\lambda$. □

Applying lemma AII.9 to the operators $Q_c(z)$ and $Q^c(z)$ we conclude that the statement of this lemma is true for $Q_c(z)$. The next step bridges the gap between this conclusion and the statement of lemma AII.8. Namely we know that the statement of lemma AII.9 remains to be true for $Q_c(z)$ if we replace in it $\mathcal{B}^{(c)}_\lambda$ by $\mathcal{B}_\lambda$. We define

$$Q_c^{new}(z) = \hat{F}_c(z) - Q_c(z)L_c(z), \qquad \text{(AII.16)}$$

Then the equation

$$R_c = F_c - R_c L_c \qquad \text{(AII.17)}$$

shows that $Q_c^{new}(z)$ satisfies (AII.10). Recalling the induction hypothesis about F_c and L_c and taking into account the conclusion above that Q_c we see that Q_c^{new} satisfies the desired properties described in lemma AII.8. □

Now we are ready to complete the study of the family $B_c(z)$.

<u>Lemma AII.10.</u> For each c with $c \subset b$, the family $B_c(z)(1 - P_c)$ is well behaved on $\mathcal{B}_\lambda$.

<u>Proof:</u> Using equation (AII.6), the equation (the parameter z is omitted)

$$R_c = F_c \sum_{k=0}^{2} (-1)^k (L_c)^k - R_c L_c^3 , \qquad \text{(AII.18)}$$

which is obtained by iterating (AII.17), and lemma AII.8 we fine

$$B_c(1-P_c)\bar{P}_c = L_c F_c^{-1} P_c F_c \sum_{k=0}^{2} (-1)^{k-1} L_c^k - L_c F_c^{-1} J_c \hat{R} Q_c L_c^3 \qquad \text{(AII.19)}$$

Since by lemma AII.6 the operators $[L^c(z)]^2$ are compact and norm-continuous for all z, the product $Q^c(L^c)^2$ is continuous in the operator norm as a function of $Q^c(z) \in L_s(\mathcal{B}_\lambda(R^c), \hat{\mathcal{B}}^c_d)$, uniformly in z on compact subsets of $\overline{\mathbb{C}^+}$ and $\overline{\mathbb{C}^-}$, by an abstract result (theorem AII.13 below). Then in virtue of lemma AII.9 (see especially equation (AII.15)), the product $Q_c(L_c)^2$ is continuous in $L(\mathcal{B}^{(c)}_\lambda, \hat{\mathcal{B}}^{(c)}_\lambda)$ as a function of $Q^c \in L_s(\mathcal{B}_\lambda(R^c), \hat{\mathcal{B}}^c_\lambda)$, uniformly on compact subsets of $\overline{\mathbb{C}^+}$ and $\overline{\mathbb{C}^-}$.

To conclude the proof we apply the induction hypothesis on L_c and $F_c^{-1}J\hat{R}$ and lemma AII.8 to the r.h.s. of (AII.19). □

Going back to the families $F_b(z)$, etc., we recall that they are built out of the B_c-operators. Then equation (AII.8) shows that we have to study the operators $B_d(z)R_\sigma(z)$ for different d and σ. This is done in the next two lemmas.

Lemma AII.11. Let $d \subset a(\sigma)$. Then the family $B_d(z)R_\sigma(z)$ has the form

$$B_dR_\sigma = R_\sigma \Lambda + \sum_{a(\delta)=d} R_\delta \Lambda_\delta ,$$

where the families $\Lambda(z)$ and $\Lambda_\delta(z)$ are well-behaved on $\mathcal{B}_\lambda$.

Proof: Because $d \subset a(\sigma)$, B_d and R_σ commute. Using (AII.8) we obtain for this case

$$R_\sigma B_d = R_\sigma B_d(\mathbb{1} - P_d) + \sum_{a(\delta)=d} R_\sigma R_\delta K_\delta$$

$$= R_\sigma B_d(\mathbb{1} - P_d) + \sum_{a(\delta)=d} (R_\sigma - R_\delta)(T_d^{a(\sigma)} - (\lambda^\sigma - \lambda^\delta))^{-1} K_\delta .$$

Condition (IE) can be written as

$$\lambda^\delta > \lambda^\sigma \quad \text{if } a(\delta) \subset a(\sigma).$$

So $0 > \lambda^\sigma - \lambda^\delta \in \rho(T_d^{s(\sigma)})$ and therefore the resolvent $(T_d^{a(\sigma)} - \lambda^\sigma + \lambda^\delta)^{-1}$ is not singular. As a result, one can easily show (using e.g. the commutator methods of section 5) that it is bounded on $\mathcal{B}_\lambda$. So the statement of the lemma follows from lemmas AII.11 and AII.5. □

Lemma AII.12. Let $d \not\subset a(\sigma)$. Then $B_d(z)R_\sigma(z)$ can be written as $\sum_{a(\delta)=d} R_\delta(z)S_\delta(z) + T(z)$, where $S_\delta(z)$ and $T(z)$ are well behaved on $\mathcal{B}_\lambda$.

Proof: We use the representation (cf. (AII.6))

$$B_d = -(\mathbb{1} + L_d)^{-1}L_d = -F_d^{-1}R_dL_d$$

$$= -\left(\sum R_\delta F_d^{-1}P_\delta + F_d^{-1}J_d\hat{R}Q_d\right)L_d , \tag{AII.20}$$

implied directly by the definition $B_d = (\mathbb{1} + L_d)^{-1} - \mathbb{1}$ and the equations $\hat{}_d + R_dL_d = F_d$ and $R_d = \sum R_\delta P_\delta + J_d\hat{R}Q_d$. We show that L_dR_σ (with $d \not\subset a(\sigma)$) is well-behaved on $\mathcal{B}_\lambda$. Then the statement of lemma AII.12 will follow from (AII.20), the induction hypothesis on $F_d(z)^{-1}$ and lemma AII.8.

In order to estimate $L_d R_\sigma$ we use representation (AI.6d) for L_d. It reduces the problem to an analysis of terms of the form $B_e R_\sigma$ with $e \subset d$. Here there are two possibilities: either $e \subset a(\sigma)$ or $e \not\subset a(\sigma)$. In the first case we apply lemma AII.11 which shows that we have to consider R_σ with the next B-operator on the left (of B_e in expression (AI.16d) for L_d). In the second case we have the same problem as we started with, only on a lower level: $e \subset d$. Here we either use the induction in d or proceed with this term as we did with $B_d R_\sigma$. In both cases the problem reduces to the lowest level, i.e. to terms of the form $B_\ell R_\sigma$ with $\ell \not\subseteq a(\sigma)$ (remember that ℓ is the partition identified with a pair ℓ). To estimate the latter terms we use again representation (AII.20) which for $d = \ell$ reduces to

$$B_\ell = (1\!\!1 - V_\ell R_\ell) V_\ell R_0 \; .$$

The term $L_\ell R_\sigma = V_\ell R_0 R_\sigma$, $\ell \not\subseteq a(\sigma)$, can be easily treated. Since $\lambda^\sigma < 0$, the last factor on the r.h.s. of

$$V_\ell R_0 R_\sigma = V_\ell (R_\sigma - R_0)(T^{a(\sigma)} - \lambda^\sigma)^{-1} \qquad \text{(AII.21)}$$

is not singular. So we can apply proposition AII.15 coupled with the commutator technique of section 5 to conclude that it is well-behaved on $\mathcal{B}_\lambda$. Iterating this procedure we arrive also at terms of the form

$$V_\ell R_{\sigma_1} R_{\sigma_2} = V_\ell (R_{\sigma_2} - R_{\sigma_1})(T^{a(\sigma_2)}_{a(\sigma_1)} - \lambda^{\sigma_2} + \lambda^{\sigma_1})^{-1} \quad \text{with} \quad a(\sigma_1) \subsetneq a(\sigma_2) \quad \text{and} \quad \ell \not\subseteq a(\sigma_1) ,$$

$a(\sigma_2)$, which we treat in the same way (remember, that $-\lambda^{\sigma_2} + \lambda^{\sigma_1} > 0$ by condition (IE)).

Summarizing, we have shown that $L_d R_\sigma$ with $d \not\subset a(\sigma)$ is well behaved on $\mathcal{B}_\lambda$. This together with equation (AII.20), the induction hypothesis on $F_d(z)$ and lemma AII.8 about Q_d completes the proof. □

Now we are fully prepared to analyze the operators $F_b(z)^{-1}$ and $L_b(z)$ (for different a satisfying $a \supseteq b$).

(i) We recall that $F_b(z)$ is representable as (AI.16) - (AI.18a). The desired properties of the disconnected part $D_b(z)$ follow directly from equation (AI.17) and lemma AII.8.

To show the desired properties of the connected part $C_b(z)$ we use representation (AI.18) with condition (AI.7b, 8b) on the summation, lemmas AII.5, AII.10-AII.12 on properties of the operators $B_c(z)$ with $c \subset b$ and the fact that $L_f R_\sigma$ with $f \not\subseteq a(\sigma)$ is well-behaved as was shown in the proof of lemma AII.12.

Condition (IE) and the fact above ensure us that all $R_\sigma(z)$ coming from different $B_{f_i}(z)$ get eventually absorbed via lemma AII.12 in one of the B_{f_i}'s, standing on the left, or in L_{f_1}.

Thus we conclude that the family $\hat{C}_b(z)$, defined by replacing all $R_{f_1}(z)$ in $C_b(z)$ by $Q_{f_1}(z)$, is well behaved from $\mathcal{B}_\lambda$ to $\hat{\mathcal{B}}_\lambda^{(b)}$. Hence, so is $\hat{F}_b(z)$.

(ii) We consider here only the $J_b\hat{R}(z)$-case. The P_β-case is simpler (this is where we use lemma AII.14 below). Using equation (AI.2) for $F_b(z)$, we get

$$F_b(z)^{-1} = \prod_{c\subset b} A_c(z)R_0(z)^{-1}, \tag{AII.22}$$

where the order in the product is opposite to that in (AI.2) (see also the remarks after equation (AI.10s)). Recalling the definition of J_b and $\hat{R}(z)$ we have

$$F_b^{-1}J_b\hat{R}\hat{u} = \sum_{\beta,a(\beta)\subset b}\ \prod_{c\subset b} A_cR_0^{-1}\psi^\beta R_\beta u_\beta, \tag{AII.23}$$

where, recall, $A_c = \mathbb{1} + L_c$. For any given $d \subset b$ the product on the right-hand side of (AII.23) can be decomposed into a sum of the term

$$A_d \prod_{c\subset d} A_cR_0^{-1} \equiv H_d - z \tag{AII.24}$$

and of monomials each of which contains at least one factor of the form L_f, $f \not\subset d$:

$$\prod_{g\subset b} A_gL_f \prod_{\psi\subset d} A_\psi R_0^{-1}. \tag{AII.25}$$

The order in the products (AII.25) is submitted to the order in the product in (AII.22). Besides, the products in (AII.25) are not necessarily taken over all partitions finer than b and d, respectively. Taking for d the decomposition a(β) in the β-term in (AII.23), we see that in the summand corresponding to (AII.24), the operator H_d-z, applied to ψ^β, cancels out the resolvent $R_\beta(z)$. In the summand corresponding to (AII.25) we commute $R_\beta(z)$ with $\prod_{\psi\subset a(\beta)} A_\psi$ and arrive at the term L_fR_β $(f \not\subset d = a(\beta))$. The operators L_fR_β with $f \not\subset a(\beta)$ were shown in the proof of lemma AII.12 to be well-behaved on $\mathcal{B}_\lambda$. Se we conclude that (ii) holds.

(iii) The proof of this part is essentially the same as the proof of the properties of the connected part, $\hat{C}_b(z)$, of $\hat{F}_b(z)$, given above in part (i), one should only use a more detailed information given in lemma AII.8 and estimates of section 5.

The proof of proposition AII.3 is completed. □

Finally we prove a simple abstract result used in the proof of lemma AII.10.

Theorem AII.13. Let $G \subset \mathbb{C}$ and X,Y and Z be Banach spaces and let A(z), $A_n(z)$, n = 1,2,..., be families of bounded operators from Y to X strongly continuous in $z \in G$ and K(z) be a family of compact operators from Z to Y, norm-continuous in $z \in G$. Then the families $A_n(z)K(z)$, n = 1,2,..., A(z)K(z) are norm continuous in $z \in G$. Moreover, if $A_n(z)$ converges strongly to A(z) uniformly in z on compacts, then $A_n(z)K(z)$ converges to A(z)K(z) in norm uniformly in z from any compact subset of G.

Proof. Let K be a compact subset of G. Since K(z) is norm continuous on K, for any $\varepsilon > 0$, K can be covered by a finite number of balls B_{i,δ_i}, centered at z_i and of radii $\delta_i>0$, such that $\|K(z)-K(w)\| \leq \varepsilon$ for any $z,w \in B_{i,\delta_i}$. Thus we have

$$\sup_{z\in K} \|(A(z_n)-A(z))K(z)\| \leq$$

$$\max_i \sup_{z\in B_{i,\delta_i}} [\|(A_n(z)-A(z))K(z_i)\| + (\|A_n(z)\| + \|A(z)\|)\varepsilon].$$

Since $A_n(z)y$ converge to $A(z)y$ in $C(K,X)$ (for any $y \in Y$), we have by the principle of uniform boundedness (see [RSI], page 81) $\sup_{z\in K} (\|A_n(z)\| + \|A(z)\|) \leq M$ for some $M < \infty$ depending on K, but independent of n.

Next, let B be the unit ball in Z. Then $K(z_i)B$ are compact sets in Y. For any $\varepsilon > 0$ there is a finite number of points y_j in Y such that the balls $B_j = \{y \in Y, \|y-y_j\| \leq \varepsilon\}$ cover $\bigcup_i K(z_i)B$. We have $\max_i \|(A_n(z)-A(z))K(z_i)\| =$

$$\sup_{x\in B} \max_i \|(A_n(z)-A(z))K(z_i)x\| \leq \max_j \max_{y\in B_j} \|(A_n(z)-A(z))y\| \leq (\max_n \|A_n(z)\| + \|A(z)\|)$$

$$+ \max_j \|(A_n(z)-A(z))y_j\| .$$

Choosing N so that $\max_{z\in K} \max_i \|(A_n(z)-A(z))y_i\| \leq \varepsilon$ for $n > N$, we conclude that for any $\varepsilon > 0$ there is N such that $\sup_{z\in K} \|(A_n(z)-A(z))K(z)\| \leq (2M+1)\varepsilon$

for $n > N$.

This completes the proof. □

Properties of the $\mathcal{B}$-spaces.

Here we describe the properties of the $\mathcal{B}$-spaces (see eqn (AII.1)) used in the study above. Note that proposition AII.15 below runs parallel to lemma 4.6 for the $\mathcal{H}$-spaces, while proposition AII.14 describes the property not possessed by the $\mathcal{H}$-spaces. The desire to have this property is exactly what caused to introduce the $\mathcal{B}$-spaces.

Proposition AII.14. The operator, I_φ, on $L^2(R)$ of the integration with $\varphi \in L^2_\delta(R^b)$, $b \subset a$, is bounded from $\mathcal{B}_{\delta,\gamma}(R)$ to $\mathcal{B}_{\delta,\gamma}(R_b)$.

Proof. The proof follows readily from the simple inequality

$$(1+|x^b|^2)^{-\delta/2}(1+|x^\ell|^2)^{-\delta/2} \leq \text{const}(1+|x^c_b|^2)^{-\delta/2} \quad \text{with} \quad c = \ell \cup b \qquad \text{(AII.26)}$$

(which results from $(1+|a+b|)^{-\delta}(1+|b|)^{-\delta} < \text{const}(1+|a|)^{-\delta}(1+|b|)^{-\delta})$. Indeed, let σ satisfy $a(\sigma) \supseteq b$, J^δ_b denote the multiplication operator by $(1+|x^b|^2)^{-\delta/2}$ and $u = J^{-\delta}_b\varphi$. Then

$$\chi_\sigma(\gamma)\langle\varphi,f\rangle = \langle\varphi,\chi_\sigma(\gamma)f\rangle = \sum_{\ell\not\subseteq a(\delta)} \langle u, J^\delta_b J^\delta_\ell h_\ell\rangle \in \sum_{c\not\subseteq a(\sigma)} J^\delta_{c(b)} L^2(R_b) \;,$$

where we have used that $\ell \not\subseteq a(\sigma)$ implies $c = b \cup \ell \not\subseteq a(\sigma)$. □

Proposition AII.15. Let I'_φ denote the operator on $L^2(R_b)$ of the tensor multiplication by $\varphi \in L^2_\delta(R^b)$: $I'_\varphi u = \varphi(x^b)u(x_b)$ (I'_φ is adjoint to I_φ) and let $J^{-\delta}_\ell V^\ell \in L^p \cap L^q(R^\nu)$, $p > \nu > q$. If $\delta > 1$ and $\ell \not\subseteq b$, then $V_\ell R_\beta(z) I'_\varphi$, $a(\beta) = b$, family of bounded operator from $\mathcal{B}_{\delta,\gamma}(R_b)$ to $\mathcal{B}_{\delta,\gamma}(R)$, $\gamma \in \kappa(\text{Re } z)$, strongly continuous as $\text{Im } z \to \pm 0$. Moreover, it is bounded in norm as

$$\|V_\ell R_\beta(z) I'_\varphi u\|_{\mathcal{B}_{\delta,\gamma}(R)} \leq C\|V_\ell\|_{L^p\cap L^q}\|\varphi\|_{L^2_\delta}\|u\|_{\mathcal{B}_{\delta,\gamma}(R)} \;,\; \gamma \in \kappa(\text{Re } z). \qquad \text{(AII.27)}$$

Proof. We begin with easy estimates:

Lemma AII.16. The operators $\chi_\alpha(\gamma)V_\ell R_\beta(z)I'_\varphi$, with $a(\alpha) \supset b \cup \ell$ and $\gamma \in \kappa(\text{Re } z)$, map $\mathcal{B}_{\delta,\gamma}(R_b)$ into $\sum_{\ell\not\subseteq a(\alpha)} J^\delta_\ell L^2(R)$ and are norm-continuous as $|\text{Im } z| \to 0$ with the norm founded by $C\|V_\ell\|_{L^p_\delta}\|\varphi\|_{L^2_\delta}$.

Proof. Let χ' be a C^∞ function such that $\chi'(t) = 1$ for $t \geq -1/5\kappa$ and $= 0$ for $t \leq -1/3\kappa$, and let $\chi'_\alpha(\gamma) = \chi'(H_\alpha-\gamma)$. Then $\chi_\alpha(\gamma) = \chi_\alpha(\gamma)\chi'_\alpha(\gamma)$.

Lemma AII.17. The operators $V_\ell \chi'_\alpha(\gamma) R_\beta(z) I'_\varphi$ with $\gamma \in \kappa(\mathrm{Re}\, z)$ are bounded from $L^2(R_b)$ to $L^2_\delta(R^b) \otimes L^2(R_b)$ (where $b = a(\beta)$) and norm-continuous as $|\mathrm{Im}\, z| \to 0$ with their norms bounded by const $\|V_\ell\|_s \|\varphi\|_{L^2_\delta}$ for any fixed $s > \max(\nu/2,2)$.

Proof. Let $\gamma \in \kappa(\lambda)$. On the subspace on which $\chi'_\alpha(\gamma) \neq 0$

$$H_\beta - \lambda = \lambda^\beta - \lambda^\alpha + T^a_b + \lambda^\alpha + T_a - \lambda \geq \lambda^\beta - \lambda^\alpha - (\tfrac{1}{2} + \tfrac{1}{3})\kappa , \qquad \text{(AII.28)}$$

where, remember, $b = a(\beta) \subsetneqq a = a(\beta)$ and $\lambda^\alpha + T_a = H_\alpha$. The condition $a(\beta) \subsetneqq a(\alpha)$ and definition (4.8) imply

$$\lambda^\beta - \lambda^\alpha - (\tfrac{1}{2} + \tfrac{1}{3})\kappa \geq \tfrac{1}{6}\kappa > 0.$$

Note here that $\kappa > 0$ follows from condition (IE). Hence the operators $\chi'_\alpha(\gamma) R_\beta(z)$, with $a(\beta) \subsetneqq a(\alpha)$ and $\gamma \in \kappa(\mathrm{Re}\, z)$, are not singular for all $z \in \mathbb{C}$ and the result follows by a standard Sobolev estimate (see e.g. example SI.2 from supplement I). □

Let now $u \in \mathcal{B}_{\delta,\gamma}(R_b)$. By the definition of $\mathcal{B}_{\delta,\gamma}(R_b)$, $\chi_\alpha(\gamma)u = \sum_{d \not\subseteq a(\alpha)} J^\delta_{d(b)} h_d$ with $h_d \in L^2(R_b)$. Since $J^\delta_{d(a)}$ commutes with all operators involved below (remember that $\ell \cup a(\beta) \subseteq a$) we have

$$\chi_\alpha V_\ell R_\beta I'_\varphi u = \sum J^\delta_{d(b)} V_\ell \chi'_\alpha R_\beta I'_\varphi h_d .$$

Since $J^\delta_{d(b)} J^\delta_b \leq J^\delta_s$ for any $s \subseteq d$ and chossing $s \not\subseteq a(\alpha)$ which is possible since $d \not\subseteq a(\alpha)$, the desired result follows by lemma AII.17. □

Now we derive the other estimates required by $\mathcal{B}_{\delta,\gamma}(R)$. Let $u \in \mathcal{B}_{\delta,\gamma}(R_b)$. Then by the definition of $\mathcal{B}_{\delta,\gamma}(R_b)$

$$V_\ell I'_\varphi R_\beta u = \sum V_\ell I'_\varphi R_\beta(z) J^\delta_{c(b)} h_c + V_\ell I'_\varphi R_\beta(z) (\mathbb{1} - \chi_\beta(\gamma)) u . \qquad \text{(AII.29)}$$

The terms on the r.h.s. are estimated in the following two lemmas.

Lemma AII.18. For any H_β- bounded operator A and any $z \in \mathbb{C}$ and $\gamma \in \kappa(\mathrm{Re}\, z)$, the operators $AR_\beta(z)\ (\mathbb{1} - \chi_\beta(\gamma))$ are bounded on $L^2(R)$ uniformly in $z \in \mathbb{C}$ and norm continuous in z.

Proof. As in the proof of lemma AII.17 we show that the operators $R_\beta(z)(\mathbb{1} - \chi_\beta(\gamma))$ are not singular if $\gamma \in \kappa(\mathrm{Re}\, z)$. This implies readily the desired result. $\mathbb{1} - \chi_\beta(\gamma) \neq 0$ only on the subspace on which $H_\beta - \gamma < -1/10\kappa$. On the other hand $\gamma \in \kappa(\mathrm{Re}\, z)$ yields that $\gamma < \mathrm{Re}\, z + 1/16\kappa$ so the former inequality implies $H_\beta - \mathrm{Re}\, z < -(1/10 - 1/16)\kappa$. □

Lemma AII.19. Let A and B be the multiplication operators by functions $\varphi \in L^p \cap L^q(R^\ell)$ and $\psi \in L^p \cap L^q(R^a_b)$, respectively. Here $p > \nu > q$ and $b \subset a$, $\#(b) = \#(a) + 1$. Then the operators $AR_\beta(z)B$, $a(\beta) = b$, are bounded on $L^2(R)$, analytic in $z \in \mathbb{C} \setminus \sigma(H_\beta)$ and strongly continuous as $z \to \sigma(H\ )$. Moreover the following estimate of their norms is true:

$$\|AR_\beta(z)B\| \leq \text{const}\ \|\varphi\|_{L^p\cap L^q}\|\psi\|_{L^p\cap L^q}. \tag{AII.30}$$

Proof. The proof is a minor modification of the proof of lemma 4.7. □

Splitting out J^δ_ℓ from V_ℓ and J^δ_b from I'_φ on the r.h.s. of (AII.29) and estimating the remaining terms by lemmas AII 18 and AII.19 we conclude that the operators $V_\ell I'_\varphi R_\beta(z)$ are bounded from $\mathcal{B}_{\delta,\gamma}(R_b)$, $\gamma \in \kappa(\text{Re } z)$, to $L^2_\delta(R^c) \otimes L^2(R_c)$ with $c = b \cup \ell$ and strongly continuous as $\text{Im } z \to \pm 0$, their norms bounded by $c\|V_\ell\|_{L^p_\delta}\|\varphi\|_{L^2_\delta}$. Moreover, for any d, $d \not\supseteq c$, $L^2_\delta(R^c) \otimes L^2(R_c) \subset \sum_{\ell \not\subseteq d} J^\delta_\ell L^2(R)$. Indeed since $d \not\supseteq c$, there exists ℓ with $\ell \subseteq c$, $\ell \not\subseteq d$. Then the inequality $J^\delta_c \leq J^\delta_\ell$ with this ℓ convinces us that the inclusion takes place. This together with lemma AII.16 completes the proof of proposition AII.15. □

Lemma AII.20. The following embeddings are true:

$$\mathcal{B}^{(c)}_\gamma(R_d) \subset \mathcal{B}_\gamma(R_d) \quad \text{and} \quad \hat{\mathcal{B}}^{(c)}_\gamma \subset \hat{\mathcal{B}}_\gamma\ .$$

Proof: The second embedding follows from the first one by definition (AII.2). We prove the first embedding. To this end we have to show that for any $f \in \mathcal{B}^{(c)}_\gamma(R_d)$ and any β with $d \subseteq a(\beta) \subset a$, we have $\chi_\beta(\gamma)f \in \sum_{e \subseteq a(\beta)} J^\delta_{e(d)} L^2(R)$.

We consider three cases of the relation of $a(\beta)$ to c. (i) $a(\beta) \subset c$. Then $\chi_\beta(\gamma)f \in \sum_{e \not\subseteq a(\beta)} J^\delta_{e(d)} L^2(R_d)$, by the definition of $\mathcal{B}^{(c)}_\gamma(R_d)$. (ii) $a(\beta) \not\subseteq c$ and $a(\beta) \not\supseteq c$. Pick σ with $a(\sigma) = a(\beta) \cap c$. Then by the definition of $\chi_\beta(\gamma)$ and since $a(\sigma) \subset a(\beta)$, $\chi_\beta(\gamma)\chi_\sigma(\gamma) = \chi_\beta(\gamma)$. Therefore taking into account that $\chi_\beta(\gamma)$ is smooth we have

$$\chi_\beta(\gamma)f \in \sum_{\substack{e \not\subseteq a(\sigma)\\ e \subset c}} J^\delta_{e(d)} L^2(R_d).$$

Moreover, since $e \not\subseteq b \cap c$ and $e \subset c$ imply $e \not\subseteq b$ (and $e \subset a$, since, remember, $c \subset a$), we have

$$\sum_{\substack{e \not\subseteq b \cap c \\ e \subseteq c}} J^{\delta}_{e(d)} L^2(R_d) \subset \sum_{\substack{e \not\subseteq b \\ e \subset a}} J^{\delta}_{e(d)} L^2(R_d),$$

which completes the consideration of this case.

(iii) $a(\beta) \supseteq c$. By the definition, $\chi_\beta(\gamma)\chi_\sigma(\gamma) = \chi_\beta(\gamma)$ for any σ with $a(\sigma) = c$. Therefore $\chi_\beta(\gamma) f \in \sum J^{\delta}_{f(c)} L^2(R_d) \cap J^{\delta}_c L^2(R_d)$. Here, recall, $(J^{\delta}_c u)(x) = (1+|x^c_d|^2)^{-\delta/2} u(x_d)$. To estimate $\chi_\rho(\gamma) f$ further we use

<u>Lemma AII.21</u>. Let $d \subseteq c$, $e \not\subseteq c$ and $f = e \cup c$. The following inequalities are true

$$J^{\delta}_{e(d)} \chi(|x^e_d| \leq \tfrac{1}{2}|x^f_c|) J^{\delta}_{f(c)} \leq \text{const}$$

and
$$J^{-\delta}_{e(d)} \chi(|x^c_d| \geq \tfrac{1}{2}|x^f_c|) J^{\delta}_c \leq \text{const}.$$

Here $\chi(\cdot)$ is the characteristic function of the set defined by the inequalities in the argument.

<u>Proof:</u> Since $d \subseteq c$ and $e \not\subseteq c$ we can write $x^e_d = y + x^f_c$, where y is a linear combination of the components of the x^c_d-vector. Then

$$(1+|y+x^f_c|^2)^{\delta/2} \chi(|x^c_d| \leq \tfrac{1}{2}|x^f_c|) \leq 2(1+|x^f_c|^2)^{\delta/2},$$

which implies the first inequality, and

$$(1+|y+x^f_c|^2)^{\delta/2} \chi(|x^c_d| \geq \tfrac{1}{2}|x^f_c|) \leq \text{const.}\ (1+|x^c_d|^2)^{\delta/2}$$

which implies the second one. □

This lemma yields

$$J^{\delta}_{f(c)} L^2(R_d) \cap J^{\delta}_c L^2(R_d) \subset J_{e(d)} L^2(R_d),$$

where e is defined by $f = e \cup c$. Since $f \not\subseteq b$ implies that $e \not\subseteq b$, this completes the consideration of the third case. □

Appendix III: Compactness of Smooth Graphs

In this appendix we prove theorem 8.8 which implies the compactness of $M(z)^2$ ("smooth graphs"). We begin with some definitions and auxilary statements.

Definition A III.1. A bounded operator, A, on $L^2(R^{a'})$ is called b-fibered, $b \subset a$, if it can be written as $Au = \int^{\oplus} a(p_b)u(p_b)dp_b$, where $a(p_b)$ is a family of bounded operators on $L^2(R^{b'})$, strongly continuous and uniformly bounded in $p_b \in (R^a_b)'$ and $u = \int^{\oplus} u(p_b)dp_b$ is a fancy way to write a vector $u \in L^2(R^{a'})$ as a vector function from $(R^a_b)'$ to $L^2(R^{b'})$, i.e. as an element of $L^2(R^{a'}_b, L^2(R^{b'}))$.

A bounded operator, A, on $L^2(R^{a'})$ is called b-compact. If it is b-fibered and its fibres, $a(p_b)$, are compact operators, norm continuous in $p_b \in (R^a_b)'$ and vanishing in norm as $|p_b|' \to \infty$.

The next two statements are related to theorem AII.13 of appendix II.

Lemma A III.2. Let A be a b-compact operator on $L^2(R^{a'})$. Let $\{b_n\}$ be a strongly converging sequence of bounded operators on $L^2(R^{b'})$. Define the operators on $L^2(R^{a'})$ by $B_n = b_n \otimes \mathbb{1}$. Then B_nA converge in the operator norm. If in addition, B_n are self-adjoint, then AB_n also converge in the operator norm.

Proof. By a special case of theorem AII.13, the sequence $\{b_na(p_b)\}$ converges in norm for each $p_b \in (R^{a'}_b)$. Hence, since $b_na(p_b)$ are equicontinuous on compacts uniformly in n ($\|b_n\|$ are uniformly bounded by the principle of uniform boundedness [RSI,p81]), it converges in norm uniformly on any compact set form $(R^{a'}_b)$. Since $\|a(p_b)\| \to 0$ as $|p_b|' \to \infty$, this implies the norm convergence of $\{b_na(p_b)\}$ uniform in $p_b \in (R^{a'}_b)$. The latter implies the norm-convergence of B_nA, which implies, if B is self-adjoint, the norm-convergence of $AB_n = (B_nA^*)^*$. □

Corollary A III.3. Let A(z) be a continuous in $z \in G$ family of b-compact operators on $L^2(R^{a'})$ and B_n be the same as in lemma A.III.2. Then $B_nA(z)$ (and, if B_n are self-adjoint, also $A(z)B_n$) converge in norm uniformly on any compact subset of z's.

Proof. Indeed, by lemma AIII.2, $B_nA(z)$ converge for each z. Since these functions are equicontinuous on compacts uniformly in n, the pointwise convergence implies the uniform convergence on compact subsets of G. □

The goal of this appendix is to prove the following result (remember that in order to fix ideas we agreed in section 8 to consider only the upper half-plane $\mathbb{C}^+$).

Proposition AIII.4. Let $M(z)$ be the operator-family described in lemma 8.7. Then for any $\varepsilon > 0$, there is a strongly-continuous family $M_1(z)$ such that

$$\| M(z) - M_1(z) \| \leq \varepsilon \tag{AIII.1}$$

and $U(\delta)M_1(z)$ and $U(-\delta)M_1(z)^*$ have analytic continuations in δ into a strip of $\mathbb{C}$ along $\mathbb{R}$ obeying $\operatorname{Im}\delta \operatorname{Im} z \leq 0$. These continuations define analytic in $z \in \bigcap_{b\subset a} \rho(H_b)$ families of compact operators, norm-continuous as $\operatorname{Im} z \downarrow 0$ $(\operatorname{Im}\delta < 0)$.

The proof of this proposition is based on a series of lemmas below.

Lemma AIII.5. $M(z)$ is a finite linear combinations of terms of the form

$$\overrightarrow{\prod_{i=1}^{k}} [U_{\ell_i} R_{\sigma_i}(z) G_{c_i}(z)], \quad z \in \bigcap \rho(H_\sigma), \tag{AIII.2}$$

where U_ℓ are multiplication operators by C_0^∞ functions of x_ℓ, $G_c(z)$ are operator-families with kernels (in the momentum representation)

$$G^c(p^c, q^c; z-\tau(p_c))\delta(p_c-q_c) \quad (\text{recall: } \tau(p_c) = \langle p_c, p_c\rangle') ,$$

where $G^c(p^c,q^c;z) \in C_0^\infty(R^{c'} \times R^{c'})$ and are infinitely and boundedly differentiable in $z \in \overline{\mathbb{C}^+}$, and the partitions satisfy

$$\forall i \ \exists\ s(i) \leq i : c_{s(i)-1} \cup \ell_{s(i)} \not\subseteq a(\sigma_i), \tag{AIII.3}$$

$$c_r \cup \ell_{r+1} \subseteq a(\sigma_i), \quad a(\sigma_r) \subsetneqq a(\sigma_i), \quad s(i) \leq r \leq i-1$$

$$\forall i \ \exists\ j(i) \geq i : c_{j(i)} \cup \ell_{j(i)+1} \not\subseteq a(\sigma_i), \tag{AIII.4}$$

$$c_{r-1} \cup \ell_r \subseteq a(\sigma_i), \quad a(\sigma_r) \subsetneqq a(\sigma_i), \quad i+1 \leq r \leq j(i)$$

$$a(\sigma_i) \subseteq c_i \subsetneqq a, \quad \bigcup_1^k (\ell_i \cup c_i) = a. \tag{AIII.5}$$

Proof. First we express $L(z)$ in the terms of the subresolvents $R_c(z)$, $c \subsetneqq a$, and potentials V_ℓ, $\ell \subseteq a$. We use representation (AI.6a) for $L(z)$ in terms of B_c, $c \subset a$, and the formulae

$$B_c = -F_c^{-1} R_c L_c \tag{AIII.6}$$

for all B_c but the one on the extreme right. For that one we use

$$B_c = -L_c F_c^{-1} R_c . \qquad \text{(AIII.7)}$$

Then we use (AI.6c) for each L_c which appears in these expressions for the B_c's and so on.

Now we must transform $L(z)$ in accordance with the proof of lemma AII.10 by replacing each R_c by

$$R_c = F_c \sum_{k=0}^{3} (-1)^k (L_c)^k - R_c L_c^4 .$$

A little contemplation of conditions (AI.7a) and (AI.8a) and the analysis in section (ii) of appendix II shows that, with or without the latter transformation, $L(z)$ is a finite sum of terms of the form

$$\vec{\prod_{i=1}} [V_{\ell_i} R_{c_i}(z)], \qquad \text{(AIII.8)}$$

where ℓ_i and c_i satisfy (AIII.3-5) with the partitions $a(\sigma_r)$ ignored and the partitions $a(\sigma_i)$ replaced by any $a_i \subseteq c_i$ (i.e. (AIII.3-5) should read $\forall i$ and $\forall a_i \subseteq c_i \; \exists \cdots$).

Finally we represent each R_c in (AIII.8) as

$$R_c = \sum_{a(\sigma)=c} R_\sigma P_\sigma + \sum_{a(\sigma)\subsetneqq c} R_\sigma J_\sigma Q_{c,\sigma} , \qquad \text{(AIII.9)}$$

where $Q_{c,\sigma}$ is the σ-component of Q_c, and replace P_σ and $J_\sigma Q_c$, by $G_{a(\sigma)}$ and $_c$, respectively. As a result we obtain terms of form (AIII.2-5). □

<u>Lemma AIII.6.</u> For any $\varepsilon > 0$ there is a strongly continuous operator-family (z) such that

$$\|(\text{AIII.2}) - T\| \leqslant \varepsilon \qquad \text{(AIII.10)}$$

nd $T(z)$ is a sum of terms each of which is obtained from (AIII.2) by replacing ach R_σ there either by $\varphi_i R_{\sigma_i} \psi_i$, where $\varphi_i \in C_0^\infty(R^{a_i})$, $a_i \supsetneqq a(\sigma_i)$, and $_i \in C_0^\infty(R^{b_i})$, $b_i \supsetneqq a(\sigma_i)$ for $i < k$ and $b_i = c_i \supseteq a(\sigma_i)$ for $i = k$, or by $_i(p_{a(\sigma_i)})$, where g_i is an entire function on $R'_{a(\sigma_i)} + iR'_{a(\sigma_i)}$, vanishing along ny tube $R'_{a(\sigma_i)} + i$ (bounded set in $R'_{a(\sigma_i)}$).

<u>Proof.</u> Consider the resolvent, R_{σ_i}, in (AIII.2) with the maximal $a(\sigma_i)$. t can be written as

$$R_{\sigma_i} = R_{\sigma_i}\chi_{\sigma_i} + R_{\sigma_i}\overline{\chi}_{\sigma_i} , \tag{AIII.11}$$

where, recall, $\chi_\sigma(\text{Re } z)$ is the cut-off operator introduced in (4.11) and $\overline{\chi}_\sigma = \mathbb{1} - \chi_\sigma$. This operator is defined so that $R_\sigma\overline{\chi}_\sigma$ and $R_\alpha\chi_\sigma$ with $a(\alpha) \subsetneq a(\sigma)$ are not singular for all z (i.e. they are bounded from H to $\mathcal{D}(T_{a(\sigma)})$ and norm-continuous in Im z and Re z). Substituting (AIII.11) into (AIII.2) splits the latter into a sum of two terms.

First we transform the χ_{σ_i}-term. Since χ_{σ_i}, taken to any positive power, commutes with U_ℓ, $\ell \subseteq a(\sigma_i)$, G_c, $c \subseteq a(\sigma_i)$, and R_α, we can write (if $i < k$)

$$G_{s(i)-1} U_{\ell_{s(i)}} \prod_{r=s(i)}^{\overrightarrow{i-1}} [R_{\sigma_r} G_{c_r} U_{\ell_{r+1}}] R_{\sigma_i}$$

$$\times \prod_{r=i+1}^{\overrightarrow{j(i)}} [G_{c_{r-1}} U_{\ell_r} R_{\sigma_r}] G_{c_{j(i)}} U_{j(i)+1} = A_i R_{\sigma_i} B_i , \tag{AIII.12}$$

where $s(i)$ and $j(i)$ are defined in (AIII.3) and (AIII.4), respectively, and

$$A_i = G_{s(i)-1} U_{\ell_{s(i)}} \prod_{r=s(i)}^{\overrightarrow{i-1}} [R_{\sigma_r} \chi^\alpha_{\sigma_i} G_{c_r} U_{\ell_{r+1}}] f_i \tag{AIII.13}$$

and

$$B_i = f_i^{-1} \prod_{r=i+1}^{\overrightarrow{j(i)}} [G_{c_{r-1}} U_{\ell_r} R_{\sigma_r} \chi^\alpha_{\sigma_i}] G_{c_{j(i)}} U_{\ell_{j(i)+1}} \tag{AIII.14}$$

with $\alpha = (j(i) - s(i))^{-1}$ and f_i, a positive function from $S(R^{a(\sigma_i)})$. For $i=k$, the terms in (AIII.12) on the right from R_{σ_i} should be replaced by just G_{c_j}. Since $R_{\sigma_r}\chi^\alpha_{\sigma_i}$ are not singular for any $\alpha > 0$ (remember that $a(\sigma_r) \subsetneq a(\sigma_i)$), the families $A_i(z)$ and $B_i(z)$ are norm-continuous in Im z and Re z and a_i- and b_i-compact, respectively. Here $a_i = \bigcup_{r=s(i)}^{i} [c_{r-1} \cup \ell_r] \cup a(\sigma_i)$ and $b_i = \bigcup_{r=i}^{j(i)} [c_r \cup \ell_{r+1}]$ for $i < k$ and $b_i = c_i$ for $i = k$.

Let $\varphi_i + \overline{\varphi}_i = \mathbb{1}$ and $\psi_i + \overline{\psi}_i = \mathbb{1}$ be partitions of unity on $L^2(R^{a_i})$ and $L^2(R^{b_i})$, respectively such that $\varphi_i \in C_0^\infty(R^{a_i})$ and $\psi_i \in C_0^\infty(R^{b_i})$. By corollary AIII.3 and the properties of A_i and B_i stated in the preceding paragraph, they can be chosen so that

$\|A_i\overline{\varphi}_i\| \leq \varepsilon$ and $\|\overline{\psi}_i B_i\| \leq \varepsilon$

Inserting $\varphi_i + \overline{\varphi}_i = 1\!\!1$ on the left of R_{σ_i} and $\psi_i + \overline{\psi}_i = 1\!\!1$, on its right we split the χ_{σ_i}-term under consideration into the sum of four terms. The terms containing either $\overline{\varphi}_i$ or $\overline{\psi}_i$ or both we throw into the ε-small basket. In the $(\varphi_i R_{\sigma_i} \psi_i)$-term we proceed to the next resolvent (i.e. the resolvent R_{σ_i} with the next largest $a(\sigma_i)$) and so on. As a result (AIII.2) is approximated as

$\|(\text{AIII.2}) - I\| \leq \varepsilon$

where I is a sum of terms each of which is obtained from (AIII.2) by replacing each $R_{\sigma_{i}}$ there by either (α) $\varphi_i R_{\sigma_i} \psi_i$ with $\varphi_i \in C_0^\infty(R^{a_i})$, $a_i \supsetneqq a(\sigma_i)$, $\psi_i \in C_0^\infty(R^{b_i})$, $b_i \supsetneqq a(\sigma_i)$ for $i < k$ and $b_i = c_i$ for $i = k$, or (β) $R_{\sigma_i} \chi'_{\sigma_i}$, where χ'_{σ_i} is either $\overline{\chi}_{\sigma_i}$ or $\chi^\alpha_{\sigma_j}$ with $\alpha > 0$ and $a(\sigma_j) \supsetneqq a(\sigma_i)$.

Now we approximate $I(z)$. In virtue of lemma AIII.10 (see remark AIII.11), given at the end of this appendix, the nonsingular factors $R_{\sigma_i}\chi'_{\sigma_i}$ can be approximate in the uniform operator topology by operators $g_i(p_{a(\sigma_i)})$, where g_i are entire functions on $R'_{a(\sigma_i)} + iR'_{a(\sigma_i)}$, vanishing along tubes $R'_{a(\sigma_i)} + i$ (bounded set in $R'_{a(\sigma_i)}$). Replacing all $R_{\sigma_i}\chi'_{\sigma_i}$ in I by approximately chosen $g_i(p_{a(\sigma_i)})$ a new operator-family $T(z)$ can be constructed which fits all the requirements of lemma AIII.6. □

<u>Lemma AIII.7.</u> Let $T(z)$ be an operator-family described in lemma AIII.6. Then $U(\delta)T(z)$ and $U(-\delta)T(z)^*$ have analytic continuations in δ into a strip of along $\mathbb{R}$ obeying $\operatorname{Im} z \leq 0$. These continuations define analytic in $z \in \bigcap_{b \subset a} \rho(H_b)$ families of compact operators, norm-continuous as $\operatorname{Im} z \downarrow 0$ ($\operatorname{Im} \delta < 0$).

<u>Proof.</u> The proof is the same for both operators mentioned in the lemma. We pick one of them, say, $U(\delta)T(z)$, for our considerations.

Let $T_1(z)$ be one of the terms constituting $T(z)$. So it is of the form

$$\Pi[U_{\ell_i} K_i G_{c_i}], \qquad \text{(AIII.15)}$$

where K_i is either $\varphi_i R_{\sigma_i} \psi_i$ or g_i. Obviously, $U(\delta)T_1(z)$ can be analytically continuated into a strip of $\overline{\mathbb{C}^-}$ adjacent to $\mathbb{R}$. Denote this continuation by $T_1^\delta(z)$. It is of the form

$$\overrightarrow{\prod_{i=1}^{k}} [U_{\ell_i}^\delta K_i^\delta G_{c_i}^\delta] \,, \tag{AIII.16}$$

where the complexly destorted operators $U_{\ell_i}^\delta$, K_i^δ and $G_{c_i}^\delta$ are defined by the condition that (AIII.16) is an analytic continuation of $U(\delta)$(AIII.15). In particular, we have in the momentum representation (here $p^2 = |p|'^2$)

$$R_{\sigma_i}^\delta(z) = (\lambda^{\sigma_i} + |p_{a(\sigma_i)}^{d_i}|'^2 + p_{d_i}^2 e^{2i\delta\varphi} - z)^{-1} \,, \tag{AIII.17}$$

where $d_i = a_i \cup [\bigcup_1^i (\ell_j \cup c_{j-1})] \not\supseteq a(\sigma_i)$ and

$$\varphi = \varphi(\sqrt{(q^{d_i})^2 + p_{d_i}^2}) \,, \tag{AIII.18}$$

$q^{d_i} \in (R^{d_i})'$ is the variable from the kernel $K(q^{d_i}, k^{d_i}; z-p_{d_i}^2)$ of the operator family, $\overrightarrow{\prod_{j=1}^{i-1}} [U_{\ell_j} K_j G_{c_j}] U_{\ell_i} \varphi_i$. A detailed form of other factors is not important.

We prove now the norm continuity of (AIII.16). To this end we assume by induction that $M_n \equiv \overrightarrow{\prod_{i-1}^{n-1}} [U_{\ell_i}^\delta K_i^\delta G_{c_i}^\delta]$ is norm-continuous as $\mathrm{Im}\, z \downarrow 0$ and prove that so is also $\overrightarrow{\prod_{i=1}^{n}} [U_{\ell_i}^\delta K_i^\delta G_{c_i}^\delta]$. For $n = 1$ we do not have the induction assumption. If $K_n = g_n$ then the result is straightforward.

Consider the case $K_n = \varphi_n R_{\sigma_n} \psi_n$. It is easy to see that $M_n \varphi_n$ is d_n-compact Since by the induction hypothesis, M_n is norm continuous as $\mathrm{Im}\, z \downarrow 0$, we have in virtue of corollary AIII.3 that for any $\varepsilon > 0$ there is a partition of unity $u + \bar{u} = \mathbb{1}$ on $L^2(R^{d_n})$ such that $u \in C_0^\infty(R^{d_n})$ and $\|M_n \varphi_n u\| \leq \varepsilon$. The next lemma shows that $uR_{\sigma_n}^\delta \psi_n^\delta$, considered on $L^2(R)$ for $n < k$ and from $\mathcal{B}_{\alpha,\gamma}(R)$ to $L^2(R)$ for $n = k$, is norm-continuous as $\mathrm{Im}\, z \downarrow 0$ which obviously completes the induction. Treating this term we might assume as well that q^{d_i} from (AIII.18) varies in a bounded region (and drop it from the consideration). Indeed, it is easy to show (R_{σ_i} are not singular for large $p^2_{a(\sigma_i)}$ and vanish when $p^2_{a(\sigma_i)} \to \infty$) that $\|\bar{\chi}_A \prod_{j=1}^{\overrightarrow{i-1}} [U_{\ell_j} K_j G_{c_j}] U_{\ell_i} \varphi_i\| \to 0$ as $A \to \infty$. Here $\bar{\chi}_A$ is the characteristic function of the set $|q^{d_i}|'^2 \geq A$ in the momentum representation. (Norm-small terms can be always ignored when one proves the norm-continuity).

Lemma AIII.8. Let f be the multiplication operator by $f \in C_0^\infty(R^d)$ and ψ, an operator with the integral kernel in the momentum representation of the form $\psi(g(p)-q^s)$, where ψ is the Fourier transform of a $C_0^\infty(\mathbb{R}^\nu)$-function ($s$ is a pair) and g is a C^∞ map of $(R_d)'$ into a sufficiently narrow tube along $(R^s)'$. Let

$$R_\alpha(z,\delta) = (\lambda^\alpha + |p^d_{a(\alpha)}|'^2 + e^{2i\delta\varphi(|p_d|')} p_d^2 - z)^{-1}$$

where $a(\alpha) \subsetneqq d$ and φ is a positive monotonically nonincreasing function on $[0,\infty)$. Then the operator-family $fR_\alpha(z,\delta)\psi$ with $s \not\subseteq a(\alpha)$ and $\delta < 0$ is norm-continuous as $\operatorname{Im} z \downarrow 0$.

Proof. Let θ be the characteristic function of the ball $p^2_{a(\alpha)} \leq A$ and $\bar{\theta} = 1 - \theta$ (again we are working in the momentum representation). If A is sufficiently large, $A > \operatorname{Re} z - \lambda^\alpha + 1$, then $R_\alpha(z,\delta)\bar{\theta}$ is not singular (it is bounded from $\mathcal{H}$ to $\mathcal{D}(T_{a(\alpha)})$ and analytic in z for $\operatorname{Re} z \leq \lambda^\alpha + A - 1$). So $fR_\alpha(z,\delta)\bar{\theta}\psi$ is also analytic in z in the same region. Now consider the family $fR_\alpha(z,\delta)\theta\psi$. It is (dUs)-fibered with Hilbert-Schmidt fibers which are Hilbert-Schmidt-norm continuous in z for $\operatorname{Re} z \leq \lambda^\alpha + A - 1$. It is especially easy to see in the case when d and s are disjoint in the sense that the indices of s form separate clusters of d (in the usual, $\delta = 0$, situation this is the most difficult case, which, besides, is responsible for the ($\delta = 0$)-operators being only strongly continuous (see lemma 4.7)). In this case the (dUs)-fibers have the following kernels in the momentum representation (for the notational convenience we set here $a(\alpha) = a_{min}$):

$$\frac{\hat{f}(p^d - q^d)\psi(g(p_d)-q^s)}{(q^d)^2 + p_d^2 e^{2i\delta\varphi(|p_d|')} - z}\, \theta(q^d, p_d)\,, \tag{AIII.19}$$

(recall that $\hat{f}$ is the Fourier transform of f). So the Hilbert-Schmidt norm of the fibres is

$$\left[\int_{q^2+p_d^2 \leq A} \left|\frac{\hat{f}(p^d-q^d)\psi(g(p_d)-q^s)}{(q^d)^2 + p_d^2 e^{2i\delta\varphi} - z}\right| dp^d dq^d dp^s dq^s\right]^{1/2}$$

$$\leq \|f\| \, |||\psi||| \left[\int_{q^2+p_d^2\leq A} \frac{dq^d dp^s}{(q^{d2}+p_d^2\cos(2\delta\varphi)-\text{Re } z)^2+|p_d|^4\sin^2(2\delta\varphi)} \right]^{\frac{1}{2}}, \quad \text{(AIII.20)}$$

where $|||\psi||| = \sup_p [\int|\psi(k+i\text{Im } g(p))|^2 dk]^{\frac{1}{2}}$. Using that $2\delta\varphi \geq \delta_1 > 0$ for $p_d^2 \leq A$ and that $p_d^2 = (p^s)^2 + (p_{dUs})^2$ due to the fact that d and s are disjoint, one shows that the last integral is bounded by a finite constant. □

This completes the proof of lemma AIII.7. □

This lemma covers all the cases except of the $i = k$ and $a(\sigma_k) = c_k$. In this case we use

Lemma AIII.9. Let f and ψ be the multiplication operators by $f \in C_0^\infty(R)$ and $\psi \in L^p \cap L^q(R^s)$, $p > \nu > q$. The operator-family $fR_\alpha(z)\psi$ with $s \not\subseteq a(\alpha)$ is uniformly bounded and norm continuous as $|\text{Im } z| \to 0$.

Proof. Let φ be a positive function from $L^p \cap L^q(R^s)$, $p > \nu > q$, and $f' = f\varphi^{-1} \in C_0^\infty(R)$ by the definition of f. By lemma 4.7, the family $\varphi R_\alpha(z)\psi$ is uniformly bounded and norm continuous as $|\text{Im } z| \to 0$. □

Lemma AIII.10. Any uniformly continuous bounded (resp. vanishing at infinity) function on $\mathbb{R}$ can be approximated by entire functions uniformly bounded (resp. vanishing at infinity) on any strip parallel to the real axis.

Proof. Let $\omega_\varepsilon(t) = (4\pi\varepsilon)^{-1} e^{-t^2/2\varepsilon}$ and define the Gauss-Weierstrass integra of f

$$(I_\varepsilon f)(x) = \int \omega_\varepsilon(x-y) f(y) dy, \quad \text{(AIII.*)}$$

where f is as described in the lemma. Then

$$\|I_\varepsilon f - f\|_\infty \to 0 \quad \text{as} \quad \varepsilon \downarrow 0. \quad \text{(AIII.21)}$$

Indeed, we have

$$|I_\varepsilon f - f| \leq \int \omega_\varepsilon(x-y)|f(y) - f(x)|dy$$

$$\leq \sup_{|y-x|\leq\delta} |f(y) - f(x)| \int_{|t|\leq\delta} \omega_\varepsilon(t)dt + 2\sup|f(x)| \int_{|t|\geq\delta} \omega_\varepsilon(t)dt.$$

Thus

$$\|I_\varepsilon f - f\|_\infty \leq \sup_{|y-x|\leq\delta} |f(y) - f(x)| + 2\|f\|_\infty (4\pi)^{-1} \int_{\delta/\varepsilon}^\infty e^{-t^2/2}dt$$

which implies (AIII.21). The functions $I_\varepsilon f$ are entire and fit the other descriptions of the statement of lemma AIII.9. □

<u>Remark AIII.11.</u> We apply this lemma to $R_{\sigma_i}\bar{\chi}_{\sigma_i} = (\lambda^{\alpha} + p^2_{a(\sigma_i)} - z)^{-1}$ $\times \; \bar{\chi}(\lambda^{\sigma_i} + p^2_{a(\sigma_i)} - \mathrm{Re}\, z)$ and $R_{\sigma_i}\chi^{\alpha}_{\sigma_j} = (\lambda^{\sigma_i} + p^2_{a(\sigma_i)} - z)^{-1}\chi(\lambda^{\sigma_j} + p^2_{a(\sigma_j)} - \mathrm{Re}\, z)^{\alpha}$ as functions of $|p_{a(\sigma_i)}|'$ and $|p_{a(\sigma_j)}|'$, respectively (recall that $a(\sigma_i) \subsetneq a(\sigma_j)$ and therefore $p^2_{a(\sigma_i)} = \left|p^{a(\sigma_j)}_{a(\sigma_i)}\right|'^2 + p^2_{a(\sigma_j)}$).

SUPPLEMENT I: Relatively Bounded, Relatively Compact and Relatively Smooth Operators

Definition SI.1. Let V and T be two closed operators on a Banach space $\mathcal{H}$. V is called T-bounded if $\mathcal{D}(V) \supset \mathcal{D}(T)$.

If V is T-bounded, then the closed graph theorem implies that V is bounded from $\mathcal{D}(T)$, equipped with the graph norm, to $\mathcal{H}$, i.e.,

$$\|Vu\| \leq a\|Tu\| + b\|u\| \quad \text{for all} \quad u \in \mathcal{D}(T) \tag{ST.1}$$

and some $a, b \geq 0$. The infimum of all a's for which (SI.1) holds is called the T-bound of V.

Example SI.2. Let $V \in L^p(\mathbb{R}^n) + L^\infty(\mathbb{R}^n)$ with $p \geq \max(\frac{n}{2}, 2)$ if $n \neq 4$ and $p > 2$ if $n = 4$. Then V is Δ-bounded. Here Δ is the Laplacian on $L^2(\mathbb{R}^n)$.

Proof. The L^∞-component of V is a bounded operator and therefore, of course, a T-bounded. To prove the relative boundedness of the L^p-component we use the Hölder inequality [RSII, St]

$$\|vu\|_2 \leq \|v\|_p \|u\|_q \quad \text{with} \quad \frac{1}{2} = \frac{1}{p} + \frac{1}{q}$$

and the embedding $H_\alpha(\mathbb{R}^n) \subset L^q(\mathbb{R}^n)$, where $\frac{1}{q} = \max(\frac{1}{2} - \frac{\alpha}{n}, 0)$, which follows from the Yound inequality and L^p-property of the integral kernel of $(1-\Delta)^{-1}$ (note that $\mathcal{D}(\Delta) = H_2(\mathbb{R}^n)$). Here $H_\alpha(\mathbb{R}^n)$ is the Sobolev space of the order α. □

Let V be a T-bounded operator and $\rho(T) \neq 0$. Then $V(T-z)^{-1}$ is a bounded operator on $\mathcal{H}$ for all $z \in \rho(T)$.

Lemma SI.3. Let T be a self-adjoint operator and V, a closed operator. Then

$$\lim_{|\mathrm{Im}z| \to \infty} \|V(T-z)^{-1}\| = \text{the T-bound of } V. \tag{SI.2}$$

Proof. We have

$$\|Vu\| \leq \|V(T+iy)^{-1}\| \, (\|Tu\| + |y| \, \|u\|)$$

for all $u \in \mathcal{D}(T)$. So (the T-bound of V) $\leq \lim_{|y| \to \infty} \|V(T+iy)^{-1}\|$. On the other hand

$$\|V(T+iy)^{-1}u\| \leq (a + b|y|^{-1}) \|u\|$$

which implies that $\lim_{|y| \to \infty} \|V(T+iy)^{-1}\| \leq$ the T-bound of V. □

<u>Theorem SI.4.</u> (Kato-Rellich). Let T be self-adjoint and V be T-bounded with the relative bound less than 1. Then $T+V$, defined with $\mathcal{D}(T+V) = \mathcal{D}(T)$, is closed. If in addition V is symmetric then $T+V$ is self-adjoint.

<u>Proof.</u> By lemma SI.3, there exist y such that $\|V(T+iy)^{-1}\| < 1$. Hence $\mathbb{1} + V(T+iy)^{-1}$ is invertible. By virtue of the equation

$$T+V+iy = (\mathbb{1} + V(T+iy)^{-1})(T+iy) \tag{SI.3}$$

the operator $T+V+iy$ with y specified above is closed as a product of a closed operator and a bounded operator which has a bounded inverse. Furthermore, by virtue of (SI.3), $T+V+iy$ is invertible for all sufficiently large y. So, if V is symmetric, the known self-adjointness criterion [RSII] implies that $T+V$ is self-adjoint. □

<u>Definition SI.5.</u> Let V and T be closed operators on $\mathcal{H}$. V is called T-compact iff $\mathcal{D}(V) \supset \mathcal{D}(T)$ and V, considered as an operator from $\mathcal{D}(T)$ to $\mathcal{H}$, is compact.

<u>Example SI.6.</u> Let $V \in L^p(\mathbb{R}^n) + (L^\infty(\mathbb{R}^n))_\varepsilon$ with $p > \max(\frac{n}{2}, 2)$ if $n \neq 4$ and $p > 2$ if $n = 4$. Here ε indicates that the L^∞-component can be taken arbitrary small. Then V is Δ-compact.

<u>Proof.</u> The proof of example SI.2 and the conditions on V imply that $V(-\Delta+\mathbb{1})^{-1}$ can be approximated in norm by operators of the form $f(x)g(i\nabla)$ with $f,g \in C_0^\infty(\mathbb{R}^n)$. Since the set of all compact operators is closed in the uniform (operator norm) topology, it suffices to show that the latter operators are compact. $f(x)g(i\nabla)$ has the kernel $f(x)\hat{g}(x-y) \in L^2(\mathbb{R}^n \times \mathbb{R}^n)$ (here $\hat{g}$ is the Fourier transform of g) and therefore is a Hilbert-Schmidt (actually, trace class) operator. □

<u>Lemma SI.7.</u> Let T be a self-adjoint operator. Any T-compact operator has the T-bound 0.

<u>Proof.</u> Since $(T+i)(T+iy)^{-1} \xrightarrow{s} 0$ as $|y| \to \infty$ for any self-adjoint operator T and $V(T+i)^{-1}$ is compact, $\|V(T+iy)^{-1}\| \to 0$ as $|y| \to \infty$. So the statement follows from lemma SI.3. □

<u>Corollary SI.8.</u> Let $V = V_1 \otimes \mathbb{1}$ and $T = T_1 \otimes \mathbb{1} + \mathbb{1} \otimes T_2$. If V_1 is T_1-compact, then V has the T-bound 0.

<u>Example SI.9.</u> Let V_i be multiplication operators on $L^2(\mathbb{R}^m)$ by $V_i(\pi_i(x))$, where $V_i \in L^{p_i}(\mathbb{R}^{n_i}) + (L^\infty(\mathbb{R}^{n_i}))_\varepsilon$ with $p_i > \max(\frac{n_i}{2}, 2)$ if $n_i \neq 4$ and $p_i > 2$ if $n_i = 4$ and π_i are linear functions from $\mathbb{R}^m$ to $\mathbb{R}^{n_i}$, $m \geq n_i$. Then $V = \Sigma V_i$ (the sum is finite) is Δ-bounded with the relative bound zero.

Proof. Let $L^2(\mathbb{R}^m) = L^2(M_{i,1}) \otimes L^2(M_{i,2})$ where $M_{i,1} = \mathrm{Ker}\,\pi_i$ and $M_{i,2} = \mathbb{R}^m \ominus M_{i,1}$. We apply corollary SI.8 to V_i and $\Delta_{i,1} \otimes 1\!\!1 + 1\!\!1 \otimes \Delta_{i,2}$, where $\Delta_{i,j}$ is the Laplacian on $L^2(M_{i,j})$. As a result, V_i have the Δ-bound 0 and so does V. □

Another very useful class of operators are Kato's relatively smooth operators defined in section 3c. Recall this definition:

Definition SI.10. (Kato). Let T be a self-adjoint operator and A, a closed operator on $\mathcal{H}$. A is called T-smooth iff for each $u \in \mathcal{H}$, $e^{iTt}u \in \mathcal{D}(A)$ for almost every $t \in \mathbb{R}$ and

$$\int_{\infty}^{\infty} \|Ae^{-iTt}u\|^2\,dt \leq C\|u\|^2 \tag{SI.4}$$

for some constant $C < \infty$.

Our task now is to prove the following statement used in the main text.

Lemma SI.11. (Kato). Let A be a closed operator and T, self-adjoint. Then A is T-smooth iff

$$A\delta_\varepsilon(T-\lambda)A^*, \text{ is bounded uniformly in } \lambda \in \mathbb{R} \text{ and } \varepsilon > 0. \tag{SI.5}$$

Here, recall $\delta_\varepsilon(T-\lambda) = \frac{1}{\pi}\,\mathrm{Im}\,R_T(\lambda+i\varepsilon)$, $\varepsilon > 0$.

Proof. The known equation

$$R_T(z) = \neg \int_0^{\pm\infty} e^{-itT}e^{izt}\,dt, \quad z \in \mathbb{C}^{\pm}, \tag{SI.6}$$

and the vector-valued version of the Plancherel theorem give

$$\pi \int_{-\infty}^{\infty} e^{-2\varepsilon|t|}\,\|Ae^{-iTt}u\|^2\,dt = \int_{-\infty}^{\infty} \|A\ \mathrm{Im}\,R_T(\lambda+i\varepsilon)u\|^2\,d\lambda\ .$$

Since $\delta_\varepsilon(T-\lambda)$ is positive we have from lemma 2.2

$$\int \|\delta_\varepsilon(T-\lambda)^{1/2}u\|^2\,d\lambda = \int \langle u, \delta_\varepsilon(T-\lambda)u\rangle\,d\lambda = \|u\|^2 .$$

Using this equality we obtain

$$\int_{-\infty}^{\infty} e^{-2\varepsilon|t|}\|Ae^{-iTt}u\|^2\,dt \leq \sup_s \|A\delta_\varepsilon(T-s)^{1/2}\|^2\ \|u\|^2\ .$$

Now the equality

$$\|A\|^2 = \|A^*\|^2 = \|A\,A^*\|$$

implies

$$\|A\delta_\varepsilon(T-\lambda)^{1/2}\|^2 = \|A\delta_\varepsilon(T-\lambda)A^*\|\ ,$$

which gives

$$\int_{-\infty}^{\infty} e^{-2\varepsilon|t|} \|Ae^{-iTt}u\|^2\, dt \lesssim \sup_{\lambda} \|A\delta_{\varepsilon}(T-\lambda)A^*\| .$$

This, by virtue of Fatou's (or B. Levy's) theorem implies that

$$\|Ae^{-iTt}u\| \in L^2(\mathbb{R}) \quad \text{and} \quad \int \|Ae^{-iTt}u\|^2\, dt \le \sup_{\lambda\ \varepsilon} \|A\delta_{\varepsilon}(T-\lambda)A^*\| \tag{SI.7}$$

The latter implies that A is T-smooth whenever (SI.5) holds.

To prove the opposite direction we notice that

$$\|A \operatorname{Im} R_T(\lambda+i\varepsilon)A^*\| = \varepsilon \|R_T(\lambda+i\varepsilon)A^*\|^2 . \tag{SI.8}$$

Using again equation (SI.6), a standard Fourier estimate and the Schwarz inequality we obtain

$$\|A R_T(\lambda+i\varepsilon)u\| \le \int_0^{\infty} e^{-\varepsilon t}\|Ae^{-iTt}u\|dt \le \varepsilon^{-1/2}[\int_0^{\infty} e^{-2\varepsilon t}\|Ae^{-iTt}u\|^2 dt]^{1/2} , \quad \varepsilon > 0.$$

This together with (SI.8) and the equation $\|B\| = \|B^*\|$ implies

$$\|A\delta_{\varepsilon}(T-\lambda)A^*\| \le \pi \sup_{\|u\|=1} [\int_{-\infty}^{\infty} \|Ve^{-iTt}u\|^2\, dt]. \tag{SI.9}$$

□

Example SI.12. (Kato). Let A be a multiplication operator by $f \in L^p \cap L^q(\mathbb{R}^n)$ with $p > n > q$. Then A is Δ-smooth.

Proof. Let T be a self-adjoint realization of the Laplacian Δ. We demonstrate (SI.5) which suffices by lemma SI.11. First we estimate

$$\|Ae^{-itT}A^*\| \le \|A^*\|_{2\to r} \|e^{-itT}\|_{r\to t'} \|A\|_{r'\to 2} .$$

By the Hölder inequality

$$\|A\|_{r'\to 2} = \|A^*\|_{2\to r} \le \|f\|_s \quad \text{with} \quad \frac{1}{r} = \frac{1}{2} + \frac{1}{s}$$

Furthermore, using the representation of e^{-itT} as an integral operator with the kernel

$$(4\pi t)^{-n/2} e^{i|x-y|^2/4t} = (4\pi t)^{-n/2} e^{ix^2/4t} e^{ix\cdot y/2t} e^{iy^2/4t}$$

and the boundedness of the Fourier transform from L^r to $L^{r'}$ for $r \le 2$, we obtain

$$\|e^{-iTt}\|_{r\to r'} \le (8\pi t)^{-\frac{n}{2}+\frac{n}{r'}}$$

Thus

$$\|Ae^{-itT}A^*\| \leq (8\pi t)^{-\frac{n}{s}} \|f\|_s^2 \quad \text{for} \quad s \geq 2.$$

Choosing $s = q$ for $|t| \leq 1$ and $s = p$ for $|t| > 1$ we get

$$\int \|Ae^{-iTt}A^*\| \, dt \leq \text{const}(\|f\|_p^2 + \|f\|_p^2).$$

Now remembering equation (SI.6) we obtain finally that $AR_T(z)A^*$ is a family of bounded operators, analytic in $z \in \mathbb{C} \setminus \overline{\mathbb{R}^+}$ and norm continuous up to $\overline{\mathbb{R}^+}$ with the bound

$$\|AR_T(z)A^*\| \leq \text{const} \, \|f\|^2_{L^p \cap L^q}, \quad p > \nu > q. \quad \square$$

SUPPLEMENT II: Balslev Combes Theorem

Balslev Combes Theorem. Let $H(\zeta)$ be the dilation analytic family associated with a quantum Hamiltonian H with real dilation analytic potentials. Then the following statements hold:

(i) $\sigma_{ess}(H(\zeta)) = \tau(H(\zeta)) + \zeta^2\overline{\mathbb{R}^+}$, where

$$\tau(H(\zeta)) = \bigcup_{1<\#(a)\leq N} \sigma_d(H^a(\zeta)),$$ the threshold set of $H(\zeta)$,

(ii) the isolated eigenvalues of $H(\zeta)$ are independent of ζ as long as they stay away from $\sigma_{ess}(H(\zeta))$. The thresholds are also locally independent of ζ,

(iii) $\sigma_p(H(\zeta)) \subset \overline{\mathbb{C}^{\pm}}$ for $\mathrm{Im}\zeta \gtrless 0$,

(iv) $\sigma_d(H(\zeta)) \cap \mathbb{R} = \sigma_p(H) \setminus \tau(H)$.

Proof. (i) Applying corollary 5.15 to $\zeta^{-2}H(\zeta) = T + \Sigma\, \zeta^{-2}V_\ell(\zeta)$ and using the Weyl criterion [K2, VH] to demonstrate the other (easy) direction we arrive at

$$\sigma_{ess}(H(\zeta)) = \bigcup_{1<\#(a)\leq N} \sigma(H_a(\zeta)) .$$

On the other hand since T_a is self-adjoint, one derives easily from the equation

$$H_a(\zeta) = H^a(\zeta) \otimes \mathbb{1} + \mathbb{1} \otimes T_a\zeta^2$$

that

$$\sigma(H_a(\zeta)) = \sigma(H^a(\zeta)) + \zeta^2\sigma(T_a).$$

Applying now the same consideration to $H^a(\zeta)$ we obtain by induction statement (i).

(ii) By the Kato analytic perturbation theorem [K2, pp 370, 375], the isolated eigenvalues of $H(\zeta)$ in a neighbourhood of any fixed ζ_0 are analytic in broken powers of ζ. On the other hand, since $H(\zeta) = U(\rho)H(e^{i\varphi})U(\rho)^{-1}$ with $\zeta = \rho e^{i\varphi}$, they are independent of $|\zeta|$. Hence they are independent of ζ as well, as long as they are away from $\sigma_{ess}(H(\zeta))$.

(iii) Assume to the contrary that $z \in \sigma_d(H(\zeta)) \cap \mathbb{C}^+$ for $\mathrm{Im}\zeta < 0$. Taking $\mathrm{Im}\zeta$ to 0 and noting that by (i), $\sigma_{ess}(H(\zeta))$ always stays in $\overline{\mathbb{C}^-}$ we conclude from

(ii) that $z \in \sigma_p(H)$. The latter contradicts the self-adjointedness of H.

(iv) Recall first that the dilation analytic (or $U(\rho)$-analytic) vectors i.e. those $f \in L^2(R)$ for which $f(\rho) = U(\rho)f$ has an analytic continuation, $f(\zeta)$, into a sector around $\overline{\mathbb{R}^+}$, form a dense subset of $L^2(R)$. Let f and g be dilation analytic vectors. Then for $z \in \overline{\mathbb{C}^+}$ and $\mathrm{Im}\zeta < 0$

$$\langle f,R(z)g\rangle = \langle f(\bar{\zeta}),R(z,\zeta)g(\zeta)\rangle , \qquad (*)$$

where $R(z,\zeta) = (H(\zeta)-z)^{-1}$. Indeed, the r.h.s. is obviously analytic in ζ, $\mathrm{Im}\zeta \lesssim 0$ for $z \in \mathbb{C}^+$ and is equal for positive ζ to the l.h.s. Hence both sides are equal for all ζ.

This equation, the spectral theorem, statements (ii), and the observation $\sigma_{ess}(H(\zeta)) \cap \mathbb{R} = \tau(H)$ (by (i), imply that $\sigma_d(H(\zeta)) \cap \mathbb{R} \subseteq \sigma_p(H) \setminus \tau(H)$. The opposite inclusion is obtained by observing that for $\lambda \in \sigma_p(H)$

$$\lim_{z\to\lambda}(\lambda-z) \langle f(\bar{\zeta}),R(z,\zeta)g(\zeta)\rangle = \langle f,Pg\rangle \neq 0,$$

where P is the eigenprojection of H associated with the eigenvalue λ. □

Spectrum of $H(\zeta)$, $\mathrm{Im}\zeta < 0$ (see Hunziker's article in [TU]).

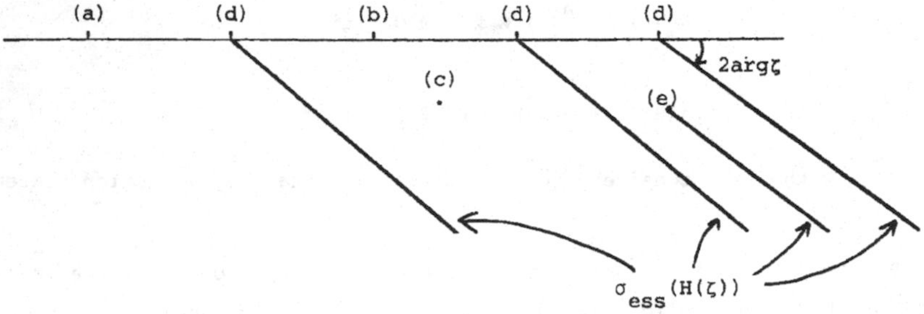

(a) = isolated eigenvalue of H
(b) = embedded eigenvalue of H (survives as $\mathrm{Im}\zeta \to 0$)
(c) = resonances eigenvalue

= set of eigenvalues of $H(\zeta)$ which can accumulate only at thresholds

(d) = real threshold
(e) = comples threshold

= $\tau(H(\zeta))$, the threshold set of $H(\zeta)$

Remarks and Reference Comments

Introduction. Excellent reviews of the mathematical quantum-mechanical scattering theory with many references can be found in [A, AJS, C5,6, N2 contains a vast number of physical discussions and applications), P3, RSIII, TU]. The following are some references to the papers on the one-body short-range scattering in addition to those already mentioned in the introduction [Ku2-4, Pr, Theo, K6,7, AS, R, Sch2].

Concerning condition (IE), recently R. Froese and I. Herbst [FH1] have shown that the many-body Schrodinger operators with potentials from a very wide class (which definitely covers all the potentials used for studying the scattering in this manuscript) have no positive eigenvalues. Previously such a result has been derived under rather stringent conditions on the potentials ([W2, B1, Sim4]; the one-body situation was rather well understood already in [K1, W1, Sim1, A12,] (see also [FHH-OH-O])).

Section 3. The two-space scattering theory was initiated by W. Hunziker [Hu2] and T. Kato [K2] (see also N.A. Shenk II [Sh] and C. Wilcox [W]) and its stationary formulation was developed by M.S. Birman et al [Bi, BeB], M. Schechter [Sch1] and D.B. Pearson [Pea] for the simple scattering systems (see book [D]), and by J. Howland [How9], T. Kato [K4] and I.M. Sigal [S4,5] in the general case. Independently and on a more formal level, the two-space theory and its applications to the multiparticle scattering were studied by C. Chandler and A.G. Gibson [ChG] and by E. Prugovecki [Pru]. Many general expressions and results of the theory are simple generalizations of the corresponding expressions and results in the one-space case as developed by J.M. Jauch [J] and W. Amrein, V. Georgescu and J.M. Jauch [AGJ].

The central theorem 3.13 was proven by T. Kato [K4] (in a somewhat different form) and independently, but later, by I.M. Sigal [S4,5]. It is motivated by the many-body results of K. Hepp [He], M. Combescure and J. Ginibre [CG], J. Howland [How9] and I.M. Sigal [S2].

Formal expressions for the scattering matrix in the two-space case were derived by C. Chandler and A.G. Gibson [ChG] and E. Prugovecki [Pru]. The structure of the three-body scattering matrix was uncovered by R. Newton [N3]. Our analysis is similar to that of K. Yajima [Yaj].

Lemma 3.21 is related to a result of T. Kato [K5].

Subsections 4a-c. (a) The kinematic description of an N-body system in terms of the spaces R, R^a and R_a with the inner product $(x,\tilde{x}) = \sum m_i x_i \cdot x_i$ is patterned on [SS]. The relatively compact potentials were introduced by J.M. Combes [C1] (see also T. Kato [K2]) and the method of dilation analytic transformation was developed by J.M. Combes [C2], J. Aguilar and J.M. Combes [AG] and E. Balslev and J.M. Combes [BC] (see also [C3,4,Sim5,TU]; a different approach was proposed by C. van Winter [vW2]). A characterization of dilation analytic potentials and vectors is given and their properties are described in [BB1,2].

(c) The existence of channel wave operators was first proven by M. Hack [Hac] who used the Cook-Kuroda method [Co, Ku1]. Hack's result was improved by many authors (see [RSIII] for references).

Sections 4d-10 and Appendices I-III. The main results of these sections, unless stated otherwise, belong to the author and can be traced to [S1-7]. Below are references to the results related to different assorted statements presented in these sections.

A more general statement than the important Kato-Iorio-O'Carroll-Combescure-Gibibre-Hagedorn lemma (lemmas 4.7 and 7.4) was proven in the momentum representation, independently, in [S2]. Lemma 5.3 generalizes a result of G. Hagedorn [H2]. Construction (5.2)-(5.4) of parametrices descends from a three-body resolvent equation of F.A. Berezin [Be].

Corollary 5.15 is usually called the difficult direction of the HVZ theorem (for W. Hunziker [Hu1], C. van Winter [vW1] and G.M. Zhislin [Z]).

An unstability of the quasibound states was first noticed by L.D. Faddeev [F] for the two-body systems and studied by G. Hagedorn [H2] for the three-body systems Different results on the instability of the embedded eigenvalues can be found in J. Howland [How1-8], W. Baumgärtel [Bau] and L. Horwitz and I.M. Sigal [HS] for the case of abstract operators and in B. Simon [Sim3] for the case of the many-body Schrödinger operators.

The references to the other results on the finiteness of the discrete spectrum can be found in [S1,8] and in Hunziker's review in [TU].

A result similar to lemma 7.14 was proven earlier by K. Hepp [He] using Schwinger's identity

$$\prod_1^s A_i^{-1} = i \int_0^\infty e^{-\sum_1^s A_i t_i} d^s t ,$$

instead of Feynman's identity used by us.

Our complex distortion techniques is related, on the one hand, to the contour deformation technique used widely in the quantum field theory and applied by

K. Hepp to the many-body quantum mechanics (see [He] and the references there) and, on the other hand, to the local distortion method due to J. Nuttall [Nu] and D. Babbitt-E. Balslev [BB3] (see also [T1, Je1,2]). Note that the latter method is related to and was motivated by the dilation transformation (or complex scalling) method [C2-6, AC, BC, Sim5, vW2] which can be called the global distortion technique. Finally, note that the contour deformation and local distortion techniques are not unrelated but the former is the source of the latter.

Supplement I. The results of this supplement belong to T. Kato, F. Rellich and J.M. Combes (see [K2, C1]).

REFERENCES

[A1] S. Agmon, Lower bounds for solutions of Schrödinger-type equations in unbounded domains, In: Proceedings of the International Congerence on Functional Analysis and Related Topics, Tokyo (1969), 216-225.

[A2] S. Agmon, Lower bounds for solutions of Schrödinger equations, Journal d'Analyse Mathematique 23, 1-25 (1970).

[A3] S. Agmon, Spectral properties of Schrödinger operators and scattering theory, Ann. Scu. Norm. Sup. Pisa Cl Sc.II, 2 (1975), 151-218.

[A4] S. Agmon, Some new results in spectral and scattering theory of differential operators on $L^2(\mathbb{R}^n)$, Seminaire Goulaic-Schwarz, 1978-79, Centre de Mathematique-Palisean, Lecture Notes.

[AC] J. Aguilar and J.M. Combes, On a class of analytic perturbations of one-body Schrödinger operators, Commun. Math. Phys. 22 (1971), 269-279.

[Am] W.O. Amrein, Some questions of non-relativistic quantum scattering theory, In: "Scattering Theory in Mathematical Physics, J. La Vita and J.P. Marchand (eds.) (1974), 97-140.

[AJS] W.O. Amrein, J.M. Jauch and K.B. Sinha, Scattering Theory in Quantum Mechanics, Reading, Mass.: Benjamin Advanced Booke Program, 1977.

[AS] P. Alsholm and G. Schmidt, Spectral and scattering theory for Schrödinger operators, Arch. Rational Mech. Anal. 40, 281-311 (1971).

[B1] E. Balslev, Absence of positive eigenvalues of Schrödinger operators, Arch. Rational Mech. Anal. 59 (1975), 343-357.

[B2] E. Balslev, Ann. Inst. Henri Poincare, 32 (1980), 125-160.

[B3] E. Balslev, Analytic scattering theory for many-body systems below the smallest three-body threshold, Commun. Math. Phys. 77 (1980), 173-210; Comments, 82 (1982), 257-260.

[BB1] D. Babbitt and E. Balslev, Dilation-analyticity and decay properties of interactions, Commun. Math. Phys. 35 (1974), 173-179.

[BB2] D. Babbitt and E. Balslev, A characterization of dilation-analytic potentials and vectors, J. of Funct. Anal. 18 (1975), 1-14.

[BB3] D. Babbitt and E. Balslev, Local distortion techniques and unitarity of the S-matrix for the 2-body problem, J. Math. Anal. Appl. 54 (1976), 316-347.

[BC] E. Balslev and J.M. Combes, Spectral properties of Schrödinger Hamiltonians with dilation analytic potentials, Commun. Math. Phys. 22 (1971), 280-294.

[Bau] W. Baumgartel, Partial resolvent and spectral concentration, Math. Nachr. 69 (1975), 107-121.

[Be] F.A. Berezin, Asymptotic behaviour of eigenfunctions in Schrödinger's equation for many particles, Dokl. Acad. Nauk USSR, 163 (1965), 795-798.

[BeB] A.L. Belopol'skii and M.S. Birman, Existence of wave operators in scattering theory for a pair of spaces, Math. USSR, Izv. 2 (1968), 1117-1130.

[Bi] M.S. Birman, A test of the existence of complete wave operators in scattering theory for a pair of spaces, In: Probl. of Math. Phys. 4, 22-26, Leningrad Univ., Leningrad (1970).

[ChG] C. Chandler and A.G. Gibson, Transition from time-dependent to time-independent multichannel quantum scattering theory, J. Math. Phys. 14 (1973), 1328-1335.

[C1] J.M. Combes, Relatively compact interactions in many particle systems, Commun. Math. Phys. 12 (1969), 283-295.

[C2] J.M. Combes, An algebraic approach to quantum scattering (1969), unpublished manuscript.

[C3] J.M. Combes, Spectral deformation techniques and applications to N-body Schrödinger operators, In: Proc. Int. Congress of Mathematicians, Vancouver (1974), 369-376.

[C4] J.M. Combes, Analytic perturbation approach to N-particle quantum systems, In: Proc. Nato Inst. on Scattering Theory, ed. J.A. La Vita and J.P. Marchand (1974), 243-272.

[C5] J.M. Combes, Scattering theory in quantum mechanics and asymptotic completeness, In: "Math. Problems in Theor.Phys", G. Dell Antonio et al (eds). Lecture Notes in Physics 80 (1978), 183-204.

[C6] J.M. Combes, Recent developments in quantum scattering theory, In: "Math. Problems in Theor. Phys.". K. Osterwalder, ed., Lecture Notes in Physics N116 (1980), 1-24.

[CG] M. Combescure (Moulin) and J. Ginibre, Hilbert space approach to the quantum mechanical three-body problem, Ann. Inst. H. Poincaré, 21 (1974), 97-145.

[Co] J. Cook, Convergence of the Moller wave matrix, J. Math. and Phys. 36 (1957), 82-87.

[D] P. Deift, Classical Scattering Theory with a Trace Condition, Princeton University Press, Princeton, N.J. ( ).

[Dev] E.B. Devies, On Enss' approach to scattering theory, Duke Math. J. 47 (1980), 171-185.

[E1] V. Enss, Asymptotic completeness for quantum mechanical potential scattering, I. Short range potentials, Commun. Math. Phys. 61 (1978), 285-291.

[E2] V. Enss, Asymptotic completeness for quantum mechanical potential scattering, II. Singular and long-range potentials, Ann. Phys. (N.Y.) 119 (1980), 117-132; Addendum, Preprint Univ. Bielefeld BI-TP 79/26.

[E3] V. Enss, Two-cluster scattering of N charged particles, Commun. Math. Phys. 65 (1979), 151-165.

[E4] V. Enss, A new method for asymptotic completeness, In: Mathematical Problems in Theor. Physics", K. Osterwalder (ed), Lecture Notes in Phys. 116, Springer (1980).

[E5] V. Enss, Geometric methods in spectral and scattering theory of Schrodinger operators, In: Rigorous Atomic and Molecular Physics, G. Vilo and A.S. Wightman (eds)., Pleum, N.Y. 1981.

[E6] V. Enss, Geometric methods in scattering theory, Acta Physica Anstriaca, Suppl. 23 (1981), 29-63.

[E7] V. Enss, Completeness of three-body quantum scattering, In: Proc. of "Bielefeld Encounters in Physics and Mathematics III", Bielefeld, Nov.30 - Dec. 4, 1981, and the series of preprints of the Ruhr-Univ., Bochum.

[F] L.D. Faddeev, Mathematical Aspects of the Three Body Problem in the Quantum Theory of Scattering, Israel Program of Scientific Translations Jerusalem (1965).

[Fr] K.O. Friedrichs, On the perturbation of continuous spectrum, Commun. Pure Appl. Math. 1 (1948), 361-406.

[FH1] R. Froese and I. Herbst, On the absence of positive eigenvalues for many-body Schrodinger operators, Preprint, Institute Mittag-Leffler, Djursholm (1982).

[FH2] R, Froese and I. Herbst, A new proof of the Mourre estimate Preprint, Institute Mittag-Leffler, Djursholm (1982).

[FHH-OH-O] R. Froese, I. Herbst, M. Hoffmann-Ostenhof and T. Hoffmann-Ostenhof, On the absence of positive eigenvalues for one-body Schrödinger operators, Preprint, Institute Mittag-Leffler, Djursholm (1982)

G] J. Ginibre, La methode "dépendent du temps" dans le probléme de la complétude asymptotique, Preprint Univ. Paris-Sud, LPTHE 80/10 (1980).

Hac] M. Hack, On the convergence to the Moller wave operators, Nuovo Cimento 9 (1958), 731-733.

H1] G. Hagedorn, A link between scattering resonances and dilation analytic resonances in few body quantum mechanics, Comm. Math. Phys. 65 (1979), 181-188.

H2] G. Hagedorn, Asymptotic completeness for classes of two, three and four particle Schrodinger operators, Trans. AMS 258 (1980), 1-75.

He] K. Hepp, On the quantum mechanical N-body problem, Helv. Phys. Acta 42 (1969), 425-458.

S] L.P. Horwitz and I.M. Sigal, On a mathematical model for non-stationary physical systems, Helv. Phys. Acta 51 (1978), 686-715.

o] L. Hörmander, Linear Partial Differential Operators. Springer-Verleg, a new edition (in preparation).

owl] J.S. Howland, J. Math. Apal. Appl. 23 (1968), 575-584.

ow2] J.S. Howland, Pac. J. Math. 29 (1969), 565-582.

ow3] J.S. Howland, Am. J. Math. 91 (1969), 1106-1126.

[How4] J.S. Howland, Trans. AMS 162 (1971), 141-156.

[How5] J.S. Howland, Bull. AMS 78 (1972), 280-283.

[How6] J.S. Howland, Regular perturbations, In: Proc. Nato Inst. on Scatt. Theory, ed. J.A. La Vita and J.M. Marchand (1974).

[How7] J.S. Howland, J. Math. Anal. Appl. 50 (1975), 415-437.

[How8] J.S. Howland, The Livsic matrix in perturbation theory, J. Math. Anal. Appl. 50 (1975), 415-437.

[How9] J.S. Howland, Abstract stationary theory of multichannel scattering, J. Funct. Anal. 22 (1976), 250-282.

[Hu1] W. Hunziker, On the spectra of Schrodinger multiparticle Hamiltonians, Helv. Phys. Acta 39 (1966), 451-462.

[Hu2] W. Hunziker, Time-dependent scattering theory for singular potentials, Helv. Phys. Acta 40 (1967), 1052-1062.

[Hu3] W. Hunziker, A remark on Enss' proof of asymptotic completeness, Preprint ETH-Zürich (1978), unpublished.

[I] T. Ikebe, Eigenfunction expansions associated with the Schrödinger operators and their application to scattering theory, Arch Rational Mech. Anal 5 (1960), 1-34.

[IOC] R.J. Iorio, Jr. and M. O'Carroll, Asymptotic completeness for multi-particle Schrödinger Hamiltonians with weak potentials. Commun. Math. Phys. 27 (1972), 137-145.

[J] J.M. Jauch, Helv. Phys. Acta 31 (1958), 127-158 and 661-684.

[Je1] A. Jensen, Local distortion technique, resonances and poles of the S-matrix, J. Math. Anal. Appl. 59 (1977), 505-513.

[J2] A. Jensen, Resonances in an abstract analytic scattering theory, Ann. Inst. H. Poincare 33 (1980), 209-223.

[K1] T. Kato, Growth properties of solutions of the reduced wave equation with a variable coefficient, Comm. Pure Appl. Math. 12 (1959), 403-425.

[K2] T. Kato, Perturbation Theory for Linear Operators, Springer-Verlag, N.Y. 1966.

[K3] T. Kato, Wave operators and similarity for some non-self-adjoint operators, Math. Annalen 162 (1966), 258-279.

[K4] T. Kato, Scattering theory with two Hilbert spaces, J. Funct. Anal., 1 (1967), 342-369.

[K5] T. Kato, Smooth operators and commutators, Studio Mathematics XXXI (1968), 535-546.

[K6] T. Kato, Some results on potential scattering, Proc. International Conf. Functional Analysis and Related Topics, Tokyo 1969, 206-215, Tokyo University Press (1970).

[K7] T. Kato, Scattering theory and perturbation of continuous spectra, Proc. International Congress Math. Nice (1970), Crauthier-Villars, Vol.1 (1971), 135-140.

[K8] T. Kato, Two-space scattering theory with applications to many-body problems, J. Fac. Sci. Univ. Tokyo 24 (1977), 503-514. (See also [TU].)

[KK1] T. Kato and S.T. Kuroda, Theory of simple scattering and eigenfunction expansions, Functional Analysis and Related Fields, 99-131, edited by F.E. Brauder, Springer, 1970.

[KK2] T. Kato and S.T. Kuroda, The abstract theory of scattering, Rocky Mountain J. Math. 1 (1971), 127-171.

[KY] H. Kitada and K. Yajima, A scattering theory for time-dependent long-range potentials, Duke Math. J. 49 (1982), 341-376.

[Ku1] S.T. Kuroda, On the existence and the unitary property of the scattering operator, Nuovo Cimento 12 (1959), 431-454.

[Ku2] S.T. Kuroda, Spectral representations and the scattering theory for Schrödinger operators, Proc. International Congress Math. Nice 1970, 441-445, Crauthier-Villars Vol. 2 (1971).

[Ku3] S.T. Kuroda, Scattering theory for differential operators I: Operator theory, J. Math. Soc. Japan 25 (1973), 75-104.

[Ku4] S.T. Kuroda, Scattering theory for differential operators II: Self-adjoint elliptic operators, J. Math. Soc. Japan 25 (1973), 222-234.

[L1] R. Lavine, Commutators and scattering theory I: Repulsive interactrons Comm. Math. Phys. 20 (1971), 301-323.

[L2] R. Lavine, Completeness of the wave operators in the repulsive N-body problem, J. Math. Phys. 14 (1973), 376-379.

[LS] M. Loss and I.M. Sigal, The three-body problem with threshold singularities, ETH-Zürich Preprint (1982).

[M] S.P. Mercuriev, Three-body Coulomb scattering, Acta Physica Austriaca, Suppl. 23 (1981), 65-110, and references given there.

[Mo1] E. Mourre, Application de la méthode de Lavine an probléme a trois corps, Ann. I. H.P. 26 (1977), 219-262.

[Mo2] E. Mourre, Link between the geometrical and the spectral transformation approaches in scattering theory, Commun. Math. Phys. 68 (1979), 91-94.

[Mo3] E. Mourre, Absence of singular spectrum for certain self-adjoint operators, Commun. Math. Phys. 78 (1981), 391-408.

[Mo4] E. Mourre, Algebraic approach to some propagation properties of the Schrödinger equation, In: Proc. of the VI Internat. Conf. on Math. Physics, W. Berlin, Aug. 11-20, 1981.

[Mo5] E. Mourre, Operateurs conjugues et proprietes de propagation, Preprint, C.N.R.S.-Luminy 1981.

[MS1] P.L. Muthuramalingam and K.B. Sinha, Asymptotic completeness in long range scattering - I, Preprint, Indian Stat. Inst.

[MS2] K.B. Sinha and P.L. Muthuramalingam, Asymptotic evolution of certain observables, Preprint, Indian Stat. Inst.

[N1] R. Newton, Scattering Theory of Waves and Particles, McGraw-Hill, N.Y., 1982.

[N2] R. Newton, Fredholm methods in the three-body problem, I, J. Math. Phys. 12 (1971), 1552-1567.

[N3] R. Newton, The three-particle S Matrix, J. Math. Phys. 15 (1974), 338-343.

[Nu] J. Nuttall, Analytic continuation of the off-energy-shell scattering amplitude, J. Math. Phys. 8 (1967), 873-877.

[Pea] D.B. Pearson, A generaliatron of the Birman trace theorem, J. Funct. Anal. 28 (1978), 182-186.

[P1] P.A. Perry, Mellin transforms and scattering theory. I. Short range potentials, Duke Math. J. 47 (1980), 187-193.

[P2] P.A. Perry, Propagation of states in dilation analytic potentials and asymptotic completeness, Commun. Math. Phys. 81 (1981), 243-260.

[P3] P.A. Perry, Scattering theory by the Enss method, to appear 1973-74.

[PSS] P. Perry, I.M. Sigal and B. Simon, Spectral analysis of N-body Schrödinger operators, Ann. of Math. 114 (1981), 519-567.

[Po] A. Ja. Povzner, The expansion of arbitrary functions in terms of eigenfunctions of the operator $-\Delta+c$, Math. Sb. 32 (1953), 109-156; AMS Translations, Series 2 60 (1967), 1-49.

[Pr] R.T. Prosser, Convergent perturbation expansion for certain wave operators, J. of Math. Physics 5 (1964), 708-713.

[Pru] E. Prugovecki, Multichannel stationary scattering theory in two-Hilbert space formulation, J. Math. Phys. 14 (1973), 957-962.

[RSI] M. Reed and B. Simon, Methods of Modern Mathematical Physics, I. Functional Analysis. Academic Press, N.Y. 1972.

[RSII] M. Reed and B. Simon, Methods of Modern Mathematical Physics, II, Fourier Analysis, Self-Adjointness. Academic Press, N.Y. 1975.

[RSIII] M. Reed and B. Simon, Methods of Modern Mathematical Physics, III. Scattering Theory. Academic Press, 1978.

[RSIV] M. Reed and B. Simon, Methods of Modern Mathematical Physics, IV. Analysis of Operators. Academic Press, N.Y., 1978.

[Schl] M. Schechter, A unified approach to scattering, J. Math. Pures Appl. (9) 57 (1974), 373-396.

[Sh] N.A. Shenk II, Eigenfunction expansions and scattering theory for the wave equation in an exterior region, Arch. Rat. Mech. Anal. 21 (1966), 121-150.

[S1] I.M. Sigal, On the discrete spectrum of the Schrödinger operators of multiparticle systems, Commun. Math. Phys. 48 (1976), 137-154.

[S2] I.M. Sigal, Mathematical foundations of quantum scattering theory for multiparticle systems, A Memoir of the AMS N209 (1978).

[S3] I.M. Sigal, On quantum mechanics of many-body systems with dilation-analytic potentials, Bull. of the AMS 84 (1978), 152-154.

[S4] I.M. Sigal, Scattering theory for multiparticle systems, I, II. ETH-Zürich (1977-78).

[S5] I.M. Sigal, Mathematical questions of quantum mechanics of many-body systems, Proc. of the Conf. on Math. Methods and Appl. of Scattering Theory, Washington, D.C. May 1979, Lecture Notes in Phys. 130 (1980) 149-158.

[S6] I.M. Sigal, Scattering theory for many-body quantum systems. Analyticity of the scattering matrix, In: "Quantum Mechanics in Mathematics, Chemistry and Physics", Plenum Publ. Corp., N.Y. (1981), 307-329.

[S7] I.M. Sigal, Mathematical theory of single-channel systems. Analyticity of scattering matrix, Trans. AMS 270 (1982), 409-437.

[S8] I.M. Sigal, Geometric methods in the quantum many-body problem. Nonexistence of very negative ions, Comm. Math. Phys. 85 (1982), 309-324.

[S9] I.M. Sigal, On a class of potentials which produce single-channel systems, unpublished.

[SS] A.G. Sigalov and I.M. Sigal, Description of the spectrum of the energy operator of quantum-mechanical systems that is invariant with respect to permutations of identical particles, Theor. and Math. Phys. 5 (1970) 990-1005.

[Sim1] B. Simon, On positive eigenvalues of one-body Schrödinger operators, Commun. Pure Appl. Math. 22 (1969), 531-538.

[Sim2] B. Simon, Phase space analysis of simple scattering systems: Extension of some work of Enss, Duke Math. J., 46 (1979), 119-168.

[Sim3] B. Simon, The theory of resonances for dilation-analytic potentials and the foundations of time dependent perturbation theory, Ann. of Math. (2) 97 (1973), 246-274.

[Sim4] B. Simon, Absence of positive eigenvalues in a class of multiparticle quantum systems, Math. Ann. 207 (1974), 133-138.

[Sim5] B. Simon, Quadratic form techniques and the Balslev-Combes theorem, Commun. Math. Phys. 27 (1972), 1-9.

[St] E. Stein, Singular Integrals and Differentiability Properties of Functions, Princeton Univ. Press, 1970.

[SW] E. Stein and G. Weiss, Introduction to Fourier Analysis on Eucleadean Spaces, Princeton Univ. Press, Princeton, N.J. 1971.

[TU] W. Thirring and P. Urban, eds., The Schrödinger Equation, Springer-Verlag, N.Y. 1977.

[Thoe] D. Theo, Eigenfunction expansions associated with Schrödinger operators in $\mathbb{R}^n$, $n \geq 4$, Arch. Rat. Mech. Anal. 26 (1967), 335-356.

[T1] L.E. Thomas, On the spectral properties of some one-particle Schrodinger Hamiltonians, Helv. Phys. Acta, 45 (1973), 1057-1065.

[T2] L.E. Thomas, Asymptotic completeness in two and three particle quantum mechanical scattering, Ann. Phys. 90 (1975), 127-165.

[VH] E. Vock and W. Hunziker, Stability of Schrödinger eigenvalue problems, Commun. Math. Phys. 83 (1982), 281-302.

W1] J. Weidmann, On the continuous spectrum of Schrödinger operators, Commun. Pure Appl. Math. 19 (1966), 107-110.

W2] J. Weidmann, The virial theorem and its applications to the spectral theory of Schrödinger operators, Bull, AMS 73 (1967), 452-456.

W] C. Wilcox, Wave operators and asymptotic solutions of wave propagation problems of classical physics, Arch. Rational Mech. Anal. 37 (1966), 37-78.

vW1] C. van Winter, Theory of finite systems of particles I and II, Danske Vid. Selsk. Mat.-Fys. Skr- 2, No.8, 10, 1-60, 1-94 (1964-1965), 1-94.

vW2] C. van Winter, Complex dynamical variables for multi-particle systems with analytic interactions, J. Math. Anal. Appl. 47 (1974), 633-670 and 48 (1974), 368-399 and the references there.

Yaf1] D.R. Yafaev, On singular spectrum of a three-body system, Mathematics USSR, Sbornic 35 (1979), 283-300.

Yaf2] D.R. Yafaev, On the proof of Enss of asymptotic completeness in potentia scattering theory, Preprint, the Leningrad branch of the Stecklov Institute.

[Yaf3] D.R. Yafaev, Asymptotic completeness for the multidimensional non-stationary Schrödinger equation, Dokl. Akad. Nauk 251 (1980), 812-816

[Yaj] K. Yajima, An abstract stationary approach to three-body scattering, J. Fac. Sci. Univ. of Tokyo, Sec. IA, 25 (1978), 109-132.

[Zh] G.M. Zhislin, Discussion of the spectrum of the Schrödinger operator for systems of many particles. Tr. Mosk. Mat. Obs. 9 (1960), 81-128.

Space Guide

Symbol	Type	Page where the space is defined
R	Finite-dimensional Real Eucledean	31
R_a, R^a	Finite-dimensional Real Eucledean	32
R_b^a	Finite-dimensional Real Eucledean	67
$\mathcal{H} = L^2(R)$	Hilbert	31
$\mathcal{H}^a = L^2(R^a)$	Hilbert	60, 67
$\mathcal{H}_\lambda$	Hilbert	22, 42
$\hat{\mathcal{H}}_\lambda$	Hilbert	16,23, 39
$\mathcal{H}_\gamma, \mathcal{H}_{\alpha,\gamma}$	Banach	36, 37
$\mathcal{B}_\gamma = \mathcal{B}_{\delta,\gamma}$	Banach	67, 87
$\mathcal{B}_\gamma^{(c)} = \mathcal{B}_{\delta,\gamma}^{(c)}$	Banach	87
$\mathcal{B}_\lambda$	Banach	87

Index of Symbols